Learning and Analytics in Intelligent Systems

Volume 23

The main aim of the series is to make available a publication of books in hard copy form and soft copy form on all aspects of learning, analytics and advanced intelligent systems and related technologies. The mentioned disciplines are strongly related and complement one another significantly. Thus, the series encourages cross-fertilization highlighting research and knowledge of common interest. The series allows a unified/integrated approach to themes and topics in these scientific disciplines which will result in significant cross-fertilization and research dissemination. To maximize dissemination of research results and knowledge in these disciplines, the series publishes edited books, monographs, handbooks, textbooks and conference proceedings.

More information about this series at http://www.springer.com/series/16172

George A. Tsihrintzis · Maria Virvou ·
Lakhmi C. Jain
Editors

Advances in Machine Learning/Deep Learning-based Technologies

Selected Papers in Honour of Professor Nikolaos G. Bourbakis – Vol. 2

Springer

Editors
George A. Tsihrintzis
Department of Informatics
University of Piraeus
Piraeus, Greece

Maria Virvou
Department of Informatics
University of Piraeus
Piraeus, Greece

Lakhmi C. Jain
KES International
Shoreham-by-Sea, UK

ISSN 2662-3447 ISSN 2662-3455 (electronic)
Learning and Analytics in Intelligent Systems
ISBN 978-3-030-76793-8 ISBN 978-3-030-76794-5 (eBook)
https://doi.org/10.1007/978-3-030-76794-5

This Springer imprint is published by the registered company Springer Nature Switzerland AG
The registered company address is: Gewerbestrasse 11, 6330 Cham, Switzerland

Foreword

Machine Learning can be considered as a part of the Artificial Intelligence field.

In 1959, Arthur Samuel [1, 2] introduced the term *Machine Learning* to refer to research efforts to develop algorithms and procedures which, when incorporated into machines, would allow them to improve their performance on specific tasks, i.e., to *learn* in ways that mimic human learning [3]. More recent efforts have been inspired by biological neural structures and have being receiving significant research attention worldwide. These approaches form a sub-area of Machine Learning, termed *Deep Learning*, and include various computing paradigms of the artificial neural network type, such as convolutional neural networks, recurrent neural networks, and deep belief networks [4].

In the six decades since publication of Samuel's nominal paper, Machine Learning, in general, and Deep Learning, in particular, have grown into one of the most active research fields worldwide. These research efforts have met with success in many technological application areas and increasingly affect many aspects of everyday life, the workplace, and human relationships [5–9]. Of course, such a broad impact also comes with risks and threats in security, privacy, safety, transparency, business, competition, the job market, fundamental rights, democracy, or even human existence itself [10–12], which are hard to ignore and care must be taken to prevent them.

Professor **Nikolaos G. Bourbakis** stands out as one of the main contributors to various applications of Machine Learning/Deep Learning throughout his long and fruitful research career at various posts. Currently, Nikolaos is a Distinguished Professor of Information & Technology and the Director of the Center of Assistive Research Technologies (CART) at Wright State University, Ohio, USA, after receiving a B.S. degree in Mathematics from the National and Kapodistrian University of Athens, Greece, a Certificate in Electrical Engineering from the University of Patras, Greece, and a Ph.D. degree in Computer Engineering and Informatics (awarded with excellence), from the Department of Computer Engineering & Informatics, University of Patras, Greece. His many achievements in Machine Learning/Deep Learning-based Technologies have been recognized via many distinctions and awards, including elevation to IEEE Fellow (1996); IEEE Computer

Society Technical Research Achievement Award; Member of the New York Academy of Sciences; Diploma of Honor in Artificial Intelligence, School of Engineering, University of Patras, Greece; ASC Outstanding Scientists & Engineers Research Award; Dr. F. Russ IEEE Biomedical Engineering Award, Dayton Ohio; Recognition Award for Outstanding Scholarly Achievements and Contributions in the field of Computer Science, University of Piraeus, Greece; IEEE EMBS-GR Award of Achievements; IEEE Computer Society 30 years ICTAI Outstanding Service & Leadership Recognition; Honorary Doctorate Degree of the University of Piraeus, Greece (2020).

Professors George A. Tsihrintzis, Maria Virvou, and Lakhmi C. Jain recently undertook a dual task. On one hand they are editing a special book in Prof. Nikolaos G. Bourbakis' honor and on the other hand they are attempting to update the relevant research communities, in computer science-related disciplines, as well as the general reader from other disciplines, on the most recent advances in Machine Learning/Deep Learning-based technological applications. They are handing to us a book consisting of 11 chapters, each of which has been written by active and recognized researchers and reports on recent research and development findings. Overall, the book is well structured as, besides an editorial note (introductory chapter), it has been further divided into five parts devoted to *Machine Learning/Deep Learning in Socializing and Entertainment* (2 chapters), *Machine Learning/Deep Learning in Education* (2 chapters), *Machine Learning/Deep Learning in Security* (2 chapters), *Machine Learning/Deep Learning in Time Series Forecasting* (2 chapters), and *Machine Learning in Video Coding and Information Extraction* (2 chapters).

Even though the area of Machine Learning/Deep Learning-based Technologies is very broad, the editors have managed to cover it impressively in terms of both breadth and depth. Undoubtedly, readers with a background in Artificial Intelligence and Computer Science will find it helpful in their researches. I am confident that interest will also be stirred among general readers who are seeking to be versed in current Machine Learning/Deep Learning-based Technologies.

I, thus, highly recommend this timely book to both the Artificial Intelligence/Computer Science researchers and the general reader.

Michalis Zervakis
Director of the Digital Image and Signal
Processing Laboratory
School of Electrical and Computer Engineering
and Vice Rector
Technical University of Crete
Crete, Greece

Further Reading

1. Arthur Samuel, Some Studies in Machine Learning Using the Game of Checkers. IBM J. 3(3), 210–229 (1959)
2. https://en.wikipedia.org/wiki/Arthur_Samuel
3. J.E. Ormrod, Human Learning, Pearson, 8th edition, ISBN-13: 978-0134893662, (2021)
4. M. Nielsen, *Neural Nets and Deep Learning* (2019), http://neuralnetworksanddeeplearning.com/
5. K. D. Foote, *A Brief History of Machine Learning* (2019) (https://www.dataversity.net/a-brief-history-of-machine-learning/)
6. History of Machine Learning, https://www.doc.ic.ac.uk/~jce317/history-machine-learning.html
7. Introduction to Neural Nets, https://www.doc.ic.ac.uk/~jce317/introduction-neural-nets.html
8. B. Grossfeld, Deep learning vs machine learning: a simple way to understand the difference, https://www.zendesk.com/blog/machine-learning-and-deep-learning/, published on January 23, 2020, last updated on October 12, 2020
9. G. A. Tsihrintzis and L. C. Jain (Eds.), Machine Learning Paradigms—Advances in Deep Learning-based Technologies, Vol. 18 in *Learning and Analytics in Intelligent Systems (LAIS)*, Springer, 2020
10. N. Bourbakis, Artificial Intelligence (AI) and its Impact to Humanity: Immortality or Last Invention, Invited Keynote Lecture, University of Piraeus, Greece, Feb. 6, (2020)
11. J. Barrat, Our Final Invention: Artificial Intelligence and the End of the Human Era, Thomas Dunne Books, October 1, 2013, ISBN-13 978-0312622374
12. Chandrasekar Vuppalapati, Democratization of Artificial Intelligence for the Future of Humanity, CRC Press, ISBN-13: 978-0367524128, January 17, (2021)

Preface

A world-recognized researcher can be honored in a variety of ways, including elevation of his professional status or various prestigious awards and distinctions. When, additionally, the same researcher has served as advisor to generations of undergraduate, graduate, and doctoral students and as mentor to faculty and colleagues, the task of appropriately honoring him becomes even harder! **Perhaps, the best way to honor this person is to ask former doctoral students, as well as colleagues and fellow researchers from around the world, to include some of their recent research results in one or more high quality volumes edited in his honor. Such an edition indicates that other researchers are pursuing and extending further what they have learned from him in research areas where he made outstanding contributions.**

Professor **Nikolaos G. Bourbakis** has been serving the fields of Artificial Intelligence (including Machine Learning/Deep Learning) and Assistive Technologies from various posts for almost fifty years now. He received a BS in Mathematics from the National and Kapodistrian University of Athens, Greece, a Certificate in Electrical Engineering from the University of Patras, Greece, and a Ph.D. in Computer Engineering and Informatics (awarded with excellence), from the Department of Computer Engineering & Informatics, University of Patras, Greece.

Dr. Bourbakis (IEEE Fellow-1996) is currently a Distinguished Professor of Information & Technology and the Director of the Center of Assistive Research Technologies (CART) at Wright State University, Ohio, USA. He is the founder and Editor-in-Chief of the International Journal on Artificial Intelligence Tools, the International Journal on Monitoring and Surveillance Technology Research (IGI-Global, Publ.), and the EAI Transactions on Bioengineering & Bioinformatics. He is also the Founder and Steering Committee Chair of several International IEEE Computer Society Conferences (namely, ICTAI, ICBIBE, ICIISA), Symposia and Workshops. He pursues research in Assistive Technologies, Applied Artificial Intelligence, Bioengineering, Information Security, and Parallel/Distributed Processing, which is funded by USA and European government and industry. He has published extensively in IEEE and International Journals and he has graduated, as the main advisor, several dozens of doctoral students. His research work has been internationally recognized and he has received several prestigious awards, including: IEEE

Computer Society Technical Research Achievement Award; Member of the New York Academy of Sciences; Diploma of Honor in AI School of Engineering, University of Patras, Greece; ASC Outstanding Scientists & Engineers Research Award; Dr. F. Russ IEEE Biomedical Engineering award, Dayton Ohio; Most Cited Article in Pattern Recognition Journal; IEEE ICTAI and ICBIBE best paper Awards; Recognition Award for Outstanding Scholarly Achievements and Contributions in the field of Computer Science, University of Piraeus, Greece; IEEE EMBS-GR Award of Achievements; IEEE Computer Society 30 years ICTAI Outstanding Service & Leadership Recognition; Honorary Doctorate degree of the University of Piraeus, Greece.

We have been collaborating with Prof. Nikolaos G. Bourbakis for very many years. Thus, we proposed and undertook with pleasure the task of editing a special book in his honor. The response from former mentees, colleagues, and fellow researchers of his has been great! Unfortunately, page limitations have forced us to limit the works to be included in the book and we apologize to those authors whose works were not included. Despite the decision not to include all proposed chapters in the book, it became apparent that not only one, but **three** volumes of the special book had to be developed, each of which would focus on different aspects of Dr. Nikolaos G. Bourbakis's research activities.

The book at hand constitutes the second volume and is devoted to **Advances in Machine Learning/Deep Learning-based Technologies**. While honoring Professor Nikolaos G. Bourbakis, this book also serves the purpose of exposing its reader to some of the most significant advances in Machine Learning/Deep Learning-based technologies. As such, the book is directed towards professors, researchers, scientists, engineers, and students in computer science-related disciplines. It is also directed towards readers who come from other disciplines and are interested in becoming versed in some of the most Advances in Machine Learning/Deep Learning-based Technologies. We hope that all of them will find it useful and inspiring in their works and researches.

We are grateful to the authors and reviewers for their excellent contributions and visionary ideas. We are also thankful to Springer for agreeing to publish this book in its **Learning and Analytics in Intelligent Systems** series. Last, but not least, we are grateful to the Springer staff for their excellent work in producing this book.

Piraeus, Greece — George A. Tsihrintzis
Piraeus, Greece — Maria Virvou
Lafayette, Indiana, USA — Lefteri Tsoukalas
Vietri, Italy — Anna Esposito
Sydney, Australia — Lakhmi C. Jain

Contents

1 Introduction to Advances in Machine Learning/Deep Learning-Based Technologies 1
George A. Tsihrintzis, Maria Virvou, and Lakhmi C. Jain
1.1 Editorial Note 1
1.2 Book Summary and Future Volumes 4
References 4

Part I Machine Learning/Deep Learning in Socializing and Entertainment

2 Semi-supervised Feature Selection Method for Fuzzy Clustering of Emotional States from Social Streams Messages 9
Ferdinando Di Martino and Sabrina Senatore
2.1 Introduction 9
2.2 The FS-EFCM Algorithm 12
2.2.1 EFCM Execution: Main Steps 13
2.2.2 Initial Parameter Setting 15
2.3 Experimental Results 16
2.3.1 Dataset 16
2.3.2 Feature Selection 17
2.3.3 FS-EFCM at Work 18
2.4 Conclusion 24
References 24

3 AI in (and for) Games 27
Kostas Karpouzis and George A. Tsatiris
3.1 Introduction 27
3.2 Game Content and Databases 28
3.3 Intelligent Game Content Generation and Selection 32
3.3.1 Generating Content for a Language Education Game 34
3.4 Conclusions 39
References 39

Part II Machine Learning/Deep Learning in Education

4 Computer-Human Mutual Training in a Virtual Laboratory Environment 47
Vasilis Zafeiropoulos and Dimitris Kalles
4.1 Introduction 47
4.1.1 Purpose and Development of the Virtual Lab 48
4.1.2 Different Playing Modes 49
4.1.3 Evaluation 50
4.2 Background and Related Work 50
4.3 Architecture of the Virtual Laboratory 51
4.3.1 Conceptual Design 51
4.3.2 State-Transition Diagrams 52
4.3.3 High Level Design 53
4.3.4 State Machine 54
4.3.5 Individual Scores 55
4.3.6 Quantization 58
4.3.7 Normalization 58
4.3.8 Composite Evaluation 59
4.3.9 Success Rate 60
4.3.10 Weighted Average 60
4.3.11 Artificial Neural Network 60
4.3.12 Penalty Points 61
4.3.13 Aggregate Score 62
4.4 Machine Learning Algorithms 63
4.4.1 Genetic Algorithm for the Weighted Average 63
4.4.2 Training the Artificial Neural Network with Back-Propagation 66
4.5 Implementation 66
4.5.1 Instruction Mode 67
4.5.2 Evaluation Mode 67
4.5.3 Computer Training Mode 68
4.5.4 Training Data Collection Sub-mode 69
4.5.5 Machine Learning Sub-mode 69
4.6 Training-Testing Process and Results 71
4.6.1 Training Data 71
4.6.2 Training and Testing on Various Data Set Groups 72
4.6.3 Genetic Algorithm Results 73
4.6.4 Artificial Neural Network Training Results 74
4.7 Conclusions 75
References 77

5 Exploiting Semi-supervised Learning in the Education Field: A Critical Survey 79
Georgios Kostopoulos and Sotiris Kotsiantis
5.1 Introduction 80
5.2 Semi-supervised Learning 81
5.3 Literature Review 82
5.3.1 Performance Prediction 83
5.3.2 Dropout Prediction 84
5.3.3 Grade Level Prediction 85
5.3.4 Grade Point Value Prediction 85
5.3.5 Other Studies 86
5.3.6 Discussion 86
5.4 The Potential of SSL in the Education Field 87
5.5 Conclusions 92
References 93

Part III Machine Learning/Deep Learning in Security

6 Survey of Machine Learning Approaches in Radiation Data Analytics Pertained to Nuclear Security 97
Miltiadis Alamaniotis and Alexander Heifetz
6.1 Introduction 97
6.2 Machine Learning Methodologies in Nuclear Security 100
6.2.1 Nuclear Signature Identification 100
6.2.2 Background Radiation Estimation 105
6.2.3 Radiation Sensor Placement 107
6.2.4 Source Localization 109
6.2.5 Anomaly Detection 110
6.3 Conclusion 112
References 112

7 AI for Cybersecurity: ML-Based Techniques for Intrusion Detection Systems 117
Dilara Gumusbas and Tulay Yildirim
7.1 Introduction 117
7.1.1 Why Does AI Pose Great Importance for Cybersecurity? 118
7.1.2 Contribution 118
7.2 ML-Based Models for Cybersecurity 119
7.2.1 K-Means 119
7.2.2 Autoencoder (AE) 121
7.2.3 Generative Adversarial Network (GAN) 122
7.2.4 Self Organizing Map 123
7.2.5 K-Nearest Neighbors (k-NN) 124
7.2.6 Bayesian Network 125
7.2.7 Decision Tree 125

7.2.8 Fuzzy Logic (Fuzzy Set Theory) 126
7.2.9 Multilayer Perceptron (MLP) 126
7.2.10 Support Vector Machine (SVM) 127
7.2.11 Ensemble Methods 128
7.2.12 Evolutionary Algorithms 129
7.2.13 Convolutional Neural Networks (CNN) 130
7.2.14 Recurrent Neural Network (RNN) 131
7.2.15 Long Short Term Memory (LSTM) 131
7.2.16 Restricted Boltzmann Machine (RBM) 132
7.2.17 Deep Belief Network (DBN) 133
7.2.18 Reinforcement Learning (RL) 133
7.3 Open Topics and Potential Directions 134
7.3.1 Novel Feature Representations 134
7.3.2 Unsupervised Learning Based Detection Systems 135
References 135

Part IV Machine Learning/Deep Learning in Time Series Forecasting

8 A Comparison of Contemporary Methods on Univariate Time Series Forecasting 143
Aikaterini Karanikola, Charalampos M. Liapis, and Sotiris Kotsiantis
8.1 Introduction 143
8.2 Related Work 146
8.3 Theoretical Background 148
8.3.1 ARIMA 148
8.3.2 Prophet 149
8.3.3 The Holt-Winters Seasonal Models 149
8.3.4 N-BEATS: Neural Basis Expansion Analysis 150
8.3.5 DeepAR 151
8.3.6 Trigonometric BATS 151
8.4 Experiments and Results 152
8.4.1 Datasets 152
8.4.2 Algorithms 152
8.4.3 Evaluation 155
8.4.4 Results 156
8.5 Conclusions 163
References 164

9 Application of Deep Learning in Recurrence Plots for Multivariate Nonlinear Time Series Forecasting 169
Sun Arthur A. Ojeda, Elmer C. Peramo, and Geoffrey A. Solano
9.1 Introduction 169
9.2 Related Work 170
9.2.1 Background on Recurrence Plots 170

9.2.2 Time Series Imaging and Convolutional Neural Networks ... 172
9.3 Time Series Nonlinearity ... 172
9.4 Time Series Imaging ... 175
9.4.1 Dimensionality Reduction ... 176
9.4.2 Optimal Parameters ... 177
9.5 Convolutional Neural Networks ... 178
9.6 Model Pipeline and Architecture ... 180
9.6.1 Architecture ... 180
9.7 Experimental Setup ... 180
9.8 Results ... 182
9.9 Conclusion ... 183
References ... 184

Part V Machine Learning in Video Coding and Information Extraction

10 A Formal and Statistical AI Tool for Complex Human Activity Recognition ... 189
Anargyros Angeleas and Nikolaos Bourbakis
10.1 Introduction ... 189
10.2 The Hybrid Framework—Formal Languages ... 191
10.3 Formal Tool and Statistical Pipeline Architecture ... 196
10.4 DATA Pipeline ... 197
10.5 Tools for Implementation ... 199
10.6 Experimentation with Datasets to Identify the Ideal Model ... 200
10.6.1 KINISIS—Single Human Activity Recognition Modeling ... 200
10.6.2 DRASIS—Change of Human Activity Recognition Modeling ... 206
10.7 Conclusions ... 213
References ... 214

11 A CU Depth Prediction Model Based on Pre-trained Convolutional Neural Network for HEVC Intra Encoding Complexity Reduction ... 217
Jiaming Li, Ming Yang, Ying Xie, and Zhigang Li
11.1 Introduction ... 218
11.2 H.265 High Efficiency Video Coding ... 220
11.2.1 Coding Tree Unit Partition ... 220
11.2.2 Rate Distortion Optimization ... 221
11.2.3 CU Partition and Image Texture Features ... 222
11.3 Proposed Methodology ... 222
11.3.1 The Hierarchical Classifier ... 223
11.3.2 The Methodology of Transfer Learning ... 223
11.3.3 Structure of Convolutional Neural Network ... 225

11.3.4 Dataset Construction 226
11.4 Experiments and Results 228
11.5 Conclusion 230
References 230

Chapter 1
Introduction to Advances in Machine Learning/Deep Learning-Based Technologies

George A. Tsihrintzis, Maria Virvou, and Lakhmi C. Jain

Abstract The field of *Machine Learning* and its sub-field of *Deep Learning* are most active areas of research in *Artificial Intelligence*, as researchers worldwide continuously develop and announce both new theoretical results and innovative applications in increasingly many and diverse other disciplines. The book at hand aims at exposing its readers to some of the most significant recent advances in Machine Learning/Deep Learning-based technologies. At the same time, the book aims at honouring Professor Nikolaos G. Bourbakis, an outstanding researcher in this area who has contributed significantly to the development of Machine Learning/Deep Learning-based technologies. As such, the book is directed towards professors, researchers, scientists, engineers and students in computer science-related disciplines. It is also directed towards readers who come from other disciplines and are interested in becoming versed in some of the most recent progress in Machine Learning/Deep Learning-based technologies. An extensive list of bibliographic references at the end of each chapter guides the readers to probe deeper into their areas of interest.

1.1 Editorial Note

The 4th Industrial Revolution is rising [1, 2], moving human civilization into a new era and restructuring human societal organization into, so-called, "Society 5.0" [3, 4]. One of the main driving forces is *Artificial Intelligence* [5], in general, and *Machine Learning* [6], in particular. The term "Machine Learning" dates back to 1959, as it was introduced in a nominal paper by Samuel [7]. Today, Machine Learning has grown

G. A. Tsihrintzis (✉) · M. Virvou
Department of Informatics, University of Piraeus, 18534 Piraeus, Greece
e-mail: geoatsi@unipi.gr

L. C. Jain
Liverpool Hope University, Liverpool, UK

University of Technology Sydney, Sydney, Australia

G. A. Tsihrintzis et al. (eds.), *Advances in Machine Learning/Deep Learning-based Technologies*, Learning and Analytics in Intelligent Systems 23,
https://doi.org/10.1007/978-3-030-76794-5_1

into a multi-disciplinary approach of very active and intense research worldwide [8–14], which aims at incorporating learning abilities into machines. More specifically, the aim of Machine Learning research and applications is to enhance machines with mechanisms, methodologies, procedures and algorithms that allow them to become better and more efficient at performing specific tasks, either on their own or with the help of a supervisor/instructor.

Within the umbrella of Machine Learning, the sub-field of *Deep Learning* (for example, see [15] for an-easy-to-follow first read on Deep Learning) includes all Machine Learning methods based on artificial neural networks, as inspired from biological neural structures. Today, Deep Learning includes multi-layered neural processing paradigms, such as convolutional neural networks, recurrent neural networks and deep belief networks [15–18].

Due to its worldwide pace of formidable growth both in new theoretical results and in achieving high scores in successful new technological application areas, the field of Machine Learning/Deep Learning has already had a significant impact on many aspects of everyday life, the workplace and human relationships and is expected to have an even wider and deeper impact in the foreseeable future [8–19].

Specifically, some of the technological application areas, where the use of Machine Learning/Deep Learning-based approaches has met with success, include the following:

I. *Machine Learning/Deep Learning in Socializing and Entertainment*
II. *Machine Learning/Deep Learning in Education*
III. *Machine Learning/Deep Learning in Security*
IV. *Machine Learning/Deep Learning in Time Series Forecasting*
V. *Machine Learning in Video Coding and Information Extraction*

The book at hand aims at updating the relevant research communities, including professors, researchers, scientists, engineers and students in computer science-related disciplines, as well as the general reader from other disciplines, on the most recent advances in Machine Learning/Deep Learning-based technological applications. At the same time, the book also aims at honouring Professor Nikolaos G. Bourbakis, an outstanding researcher and educator, who has conducted leading research in this field and has inspired, advised and mentored tens of students, fellow researchers and colleagues.

More specifically, the book consists of an editorial chapter (Chap. 1) and an additional ten (10) chapters. All chapters in the book were invited from authors who work in the corresponding chapter theme and are recognized for their significant research contributions. In more detail, the chapters in the book are organized into five parts, as follows:

The ***first part*** of the book consists of two chapters devoted to ***Machine Learning/Deep Learning in Socializing and Entertainment.***

Specifically, Chap. 2, by F. Di Martino and S. Senatore, is entitled "*Semi-Supervised Feature Selection Method for Fuzzy Clustering of Emotional States from Social Streams Messages.*" The authors propose a new method based on a fuzzy clustering algorithm that takes into account human suggestions for feature selection to capture the user mood in decision-making processes.

Chapter 3, by K. Karpouzis and G. Tsatiris, is entitled "*AI in (and for) Games.*" The authors discuss some of the most common and widely accepted uses of Artificial Intelligence/Machine Learning algorithms in games and how intelligent systems can benefit from those, elaborating on estimating player experience based on expressivity and performance, and on generating proper and interesting content for a language learning game.

The ***second part*** of the book consists of two chapters devoted to ***Machine Learning/Deep Learning in Education.***

Specifically, Chap. 4, by V. Zafeiropoulos and D. Kalles, is entitled "*Computer-Human Mutual Training in a Virtual Laboratory Environment.*" The authors discuss recent developments of Onlabs, an interactive 3D virtual lab developed at the Hellenic Open University, and assess its performance via two separate machine learning techniques, namely a genetic algorithm and back-propagation on an artificial neural network.

Chapter 5, by G. Kostopoulos and S. Kotsiantis, is entitled "*Exploiting Semi-supervised Learning in the Education Field: A Critical Survey.*" The authors provide a comprehensive review of the applications of Semi-Supervised Learning in the fields of Educational Data Mining and Learning Analytics. Their review indicates that Semi-Supervised Learning constitutes a very effective tool for both early and accurate prognosis of student learning outcomes when compared to traditional supervised methods.

The ***third part*** of the book consists of two chapters devoted to ***Machine Learning/Deep Learning in Security.***

Specifically, Chap. 6, by M. Alamaniotis and A. Heifetz, is entitled "*Survey of Machine Learning Approaches in Radiation Data Analytics pertained to Nuclear Security.*" The authors provide a comprehensive survey and discussion of Machine Learning and Data Analytics methods pertaining to nuclear security and also discuss further trends and how the data analytics can further enhance nuclear security by effectively analyzing radiation data.

Chapter 7, by D. Gumusbas and T. Yildirim, is entitled "*AI for Cybersecurity: ML-Based Techniques for Intrusion Detection Systems.*" The authors discuss problems in cybersecurity and their potential Machine Learning-based solutions and point to open avenues of future research in this area.

The ***fourth part*** of the book consists of two chapters devoted to ***Machine Learning/Deep Learning in Time Series Forecasting***.

Specifically, Chap. 8, by A. Karanikola, C. M. Liapis and S. Kotsiantis, is entitled "*A Comparison of Contemporary Methods on Univariate Time Series Forecasting.*" The authors compare the performance of several contemporary forecasting models that are considered state of the art, including Autoregressive Integrated Moving Average (ARIMA), Neural Basis Expansion Analysis (NBEATS), Probabilistic Time Series Modeling, Deep Learning-based models and other.

Chapter 9, by S. A. A. Ojeda, E. C. Peramo, and G. A. Solano, is entitled "*Application of Deep Learning in Recurrence Plots for Multivariate Nonlinear Time*

Series Forecasting." The authors present a framework for multivariate nonlinear time series forecasting that utilizes phase space representations and deep learning.

Finally, the ***fifth part*** of the book consists of two chapters devoted to ***Machine Learning in Video Coding and Information Extraction.***

Specifically, Chap. 10, by A. Angeleas and N. G. Bourbakis, is entitled "*Formal and Statistical AI Tool for Complex Human Activity Recognition.*" The authors present a novel end-to-end Machine Learning-based tool for complex human activity recognition and behavioral interpretation, backed by formal and statistical information.

Finally, Chap. 11, by J. Li, M. Yang, Y. X. and Z. Li, is entitled "*A CU Depth Prediction Model Based on Pre-trained Convolutional Neural Network for HEVC Intra Encoding Complexity Reduction.*" The authors' work is in the area of High Efficiency Video Coding and proposes a hierarchical Coding Unit depth prediction model based on a pre-trained convolutional neural network to predict the Coding Tree Unit split pattern based on the image block.

1.2 Book Summary and Future Volumes

In this book, we have presented some significant advances in Machine Learning/Deep Learning-based technologies, while honouring Professor Nikolaos G. Bourbakis for his research contributions to this discipline. The book is directed towards professors, researchers, scientists, engineers and students in computer science-related disciplines. It is also directed towards readers who come from other disciplines and are interested in becoming versed in some of the most recent advances in these active technological areas. We hope that all of them will find the book useful and inspiring in their works and researches.

The book has also come as a seventh volume, following six previous volumes by the Editors devoted to aspects of various Machine Learning Paradigms [8–10, 12, 14, 19]. As societal demand continues to pose challenging problems, which require ever more efficient tools, methodologies, systems and computer science-based technologies to be devised to address them, the readers may expect that additional related volumes will appear in the future.

References

1. J. Toonders, Data is the new oil of the digital economy. Wired. https://www.wired.com/insights/2014/07/data-new-oil-digital-economy/
2. K. Schwabd, The Fourth Industrial Revolution—what it means and how to respond. Foreign Affairs. https://www.foreignaffairs.com/articles/2015-12-12/fourth-industrial-revolution. Accessed 12 Dec 2015
3. From Industry 4.0 to Society 5.0: the big societal transformation plan of Japan, https://www.i-scoop.eu/industry-4-0/society-5-0/.

4. Society 5.0, https://www8.cao.go.jp/cstp/english/society5_0/index.html
5. E. Rich, K. Knight, S.B. Nair, *Artificial Intelligence*, 3rd edn. (Tata McGraw-Hill Publishing Company, 2010)
6. J. Watt, R. Borhani, A.K. Katsaggelos, *Machine Learning Refined—Foundations Algorithms and Applications,* 2nd edn. (Cambridge University Press, 2020)
7. A. Samuel, Some studies in machine learning using the game of checkers. IBM J. **3**(3), 210–229 (1959)
8. A.S. Lampropoulos, G.A. Tsihrintzis, Machine learning paradigms—applications in recommender systems, in *Intelligent Systems Reference Library Book Series*, vol. 92 (Springer, 2015)
9. D.N. Sotiropoulos, G.A. Tsihrintzis, Machine learning paradigms—artificial immune systems and their application in software personalization, in *Intelligent Systems Reference Library Book Series*, vol. 118 (Springer, 2017)
10. G.A. Tsihrintzis, D.N. Sotiropoulos, L.C. Jain (eds.), Machine learning paradigms—advances in data analytics, in *Intelligent Systems Reference Library Book Series*, vol. 149 (Springer, 2018)
11. A.E. Hassanien (ed.), Machine learning paradigms: theory and application, in *Studies in Computational Intelligence Book Series*, vol. 801 (Springer, 2019)
12. G.A. Tsihrintzis, M. Virvou, E. Sakkopoulos, L.C. Jain (eds.), Machine learning paradigms—applications of learning and analytics in intelligent systems, in *Learning and Analytics in Intelligent Systems Book Series*, vol. 1 (Springer, 2019)
13. J.K. Mandal, S. Mukhopadhyay, P. Dutta, K. Dasgupta (eds.), Algorithms in machine learning paradigms, in *Studies in Computational Intelligence Book Series*, vol. 870 (Springer, 2020)
14. M. Virvou, E. Alepis, G.A. Tsihrintzis, L.C. Jain (eds.), Machine learning paradigms—advances in learning analytics, in *Intelligent Systems Reference Library Book Series*, vol. 158 (Springer, 2020)
15. J. Patterson, A. Gibson *Deep Learning—A Practitioner's Approach* (O' Reilly, 2017)
16. Y. Bengio, A. Courville, P. Vincent, Representation learning: a review and new perspectives. IEEE Trans. Pattern Anal. Mach. Intell. **35**(8), 1798–1828 (2013)
17. J. Schmidhuber, Deep learning in neural networks: an overview. Neural Netw. **61**, 85–117 (2015)
18. Y. Bengio, Y. LeCun, G. Hinton, Deep learning. Nature **521**(7553), 436–444 (2015)
19. G.A. Tsihrintzis, L.C. Jain (eds.), Machine learning paradigms—advances in deep learning-based technological applications, in *Learning and Analytics in Intelligent Systems Book Series*, vol. 18 (Springer, 2020)

Part I
Machine Learning/Deep Learning in Socializing and Entertainment

Chapter 2
Semi-supervised Feature Selection Method for Fuzzy Clustering of Emotional States from Social Streams Messages

Ferdinando Di Martino and Sabrina Senatore

Abstract Capturing the text content, especially when it reflects the human emotional states and feelings, is crucial in every decision-making process: from the item purchase to the marketing campaign, the user mood is becoming an essential peculiarity to always monitoring. This work proposed a new method based on a fuzzy clustering algorithm that takes into account human suggestions for feature selection. The method exploits two fuzzy indices, namely, the *feature relevance* that is initially provided by the human expertise and the *feature incidence* on a specific cluster. The Extended Fuzzy C-Means (EFCM) clustering is used to balance the two "dueling" indexes; a t-norm operator-based feature importance index enables the appropriate feature set selection. Experimental results on social message streams show the method's effectiveness in supporting those emotions the human considers relevant in the textual context.

2.1 Introduction

Nowadays, high throughput technologies routinely produce large data that are recorded and stored for analytics purposes. In the Social Web particularly, the continuous user-generated content needs to be arranged appropriately to accelerate text analysis and information retrieval tasks. Data often contain irrelevant and redundant features, with a high level of noise.

F. Di Martino (✉)
Dipartimento di Architettura, Università degli Studi di Napoli Federico II, Via Toledo 402, 80134 Napoli, Italy
e-mail: fdimarti@unina.it

S. Senatore
Dipartimento di Ingegneria dell'Informazione ed Elettrica e Matematica applicata, Università degli Studi di Salerno, Via Giovanni Paolo II, 132, 84084 Fisciano, Salerno, Italy
e-mail: ssenatore@unisa.it

G. A. Tsihrintzis et al. (eds.), *Advances in Machine Learning/Deep Learning-based Technologies*, Learning and Analytics in Intelligent Systems 23,
https://doi.org/10.1007/978-3-030-76794-5_2

Especially in the classification task, large feature vectors could significantly slow down the process, and, even though such vectors are expected to have more discriminating power, practically, they often produce models that do not reflect a quite generalized data representation.

This problem is quite evident in the processing of textual information, such as papers, websites, reviews, twitters, or snippets. The expressiveness of natural language emphasizes the difficulty to discriminate appropriate features that support accurately the classification methods. On the other hand, the increasing volume of opinionated data disseminated on the Web needs enhanced approaches to analyze and process data in an efficient way, capturing the actual meaning behind the text. The natural language indeed is imprecise and ambiguous and, in general, the text is composed of a loosely structured sequence of words and symbols that can support humans in capturing the actual meaning of the sentences but, this activity is quite complex for the computational systems that could not infer a right context for a group of words. These issues can be amplified if text mining activities are targeted at capturing the emotions and the sentiments from opinions [20].

Analyzing user-generated content in social media to capture people's emotions and understand public attitude and mood is a crucial task for market analysis, business, political consensus study. Consumers can influence other users' consumption activities: an opinion, a comment, a reaction quickly reach global audiences who share similar interests in a product or brand. Text Mining is a complex activity that needs to discover relevant information from a large collection of textual information, which is often unstructured, redundant, and duplicated.

Feature selection becomes a mandatory preprocessing phase to reduce the dimensionality and eliminate the duplication and unwanted features in the data.

Many feature selection algorithms have been developed in literature [7], often studied as optimization problems [9]. In Information Retrieval (IR) approaches, the Bag-of-Words (BoW) is the most known vector space model used to represent documents. It is filled by word frequencies over a fixed dictionary. The feature selection methods remove the lowest ranking terms based on a scoring function, such as term occurrences [16] and frequencies, TF-IDF [17], as well as mutual information (MI) and chi-squared ranking ($\chi 2$) [12].

Selecting the highest-ranked terms does not guarantee to get the most relevant feature, especially in the text mining tasks, where polysemy and synonymy can affect the classification tasks: redundant features do not contribute to adding new information to describing the concept, as well as irrelevant features can simply add noise to the mining process [20]. On the other hand, the reduction of the feature set impacts the size of the data space, and therefore also decreases the complexity of the classification and prediction problems.

In addition to the traditional Information Retrieval and Text Mining approaches, which are mainly based on a preliminary feature set definition, many approaches in the literature aim at seeking latent semantics from the data space to face polysemy, synonyms, homonyms, phrases dependencies issues. Latent models [1] are useful to identify the semantic concepts in text documents and uncover the latent semantic structure embedded in document collections. In [5] latent models help discriminating emotions from textual space: the projection of words expressing sentiments and emotions in the same space topology provides effective contexts to overcome linguistic ambiguity in the natural language. Sentiment Analysis has been extensively studied in recent years [11–19], often focusing on natural language processing techniques to face all the issues related to understanding the written language targeted at interpreting human moods.

In [1], the Latent Dirichlet Allocation (LDA) model is used to extract latent topics and is combined with a Bayesian approach to extract concepts to associate with the topics. Some approaches [8–21] adopt external resources such as WordNet [22] as well as dictionaries, thesauri, and knowledge bases to discriminate sense and the context of terms in sentences.

Classification in natural language processing tasks finds in Deep Learning techniques [14] compelling methods to capture the complexity of language, overcoming problems, such as the curse of dimensionality, since the linguistic text was represented with sparse matrices (high-dimensional features). With the recent popularity of word embeddings, neural-based approaches exhibit good performance compared to more traditional machine learning models. Empirical evidence shows that discovering linguistic patterns remain an open issue in language understanding, due to the complexity of natural language, that through metaphors, rhetoric, figurative expressions make ineffective known automatic models for feature extraction and selection.

This paper presents a novel feature selection method applied to fuzzy clustering. The algorithm is called FS-EFCM (Future Selection on Extended Fuzzy C-Means) and extends the EFCM algorithm [10]. The algorithm takes external scores into account as additional parameters for the initial configuration. The idea is to allow human suggestion in the discrimination of relevant features, viz., the features that are crucial to describe the domain of interest the features are from, and then to affect the clustering process into discarding irrelevant and noise information.

Thus, experts can provide their relevance values (weights) to each feature based on their expertise and knowledge of the reference domain.

During the EFCM execution, some features could be discarded based on the expert feature relevance selection. The features are also evaluated with respect to their impact, i.e., the affecting on the formation of the clusters. Both feature relevance and incidence are monitored during the FS-EFCM execution, to evaluate which features are crucial for both the clustering performance and the experts.

The remainder of the paper is organized as follows. Section 2.2 introduces the proposed FS-EFCM algorithm: a general overview of the algorithm is presented firstly, then the main steps and the pseudocode describe the algorithm in detail. Additional investigation on the parameter setting configuration is also provided. Section 2.3 is devoted to the experimental results. A dataset composed of tweet streams is analyzed to classify the tweet trends by capturing the sentiments and emotions from text analysis. The experiments show the effectiveness of the proposed method. Finally, conclusions will be given in the last section.

2.2 The FS-EFCM Algorithm

The FS-EFCM algorithm accomplishes data fuzzy clustering by introducing a feature selection method that filters irrelevant features out. It is from an extension of the EFCM [10], a partitive fuzzy clustering algorithm that, based on the well-known FCM algorithm [4, 3], overcomes its drawbacks, such as a priori choice of the number of clusters and the sensibility to the presence of noise and outliers. Additionally, EFCM is robust to the partition initialization and does not require to validate the produced clustering partitioning over several random initializations.

Figure 2.1 sketches the main steps of the FS-EFCM algorithm, showing the EFCM module embedded. During the EFCM runs, some features are candidates to be discarded, since they are not meaningful in the feature set and, at the same time, they do not affect the cluster formation. The EFCM is then re-executed until the *stability condition* is not satisfied.

The algorithm takes as input the data collection and the expert-driven scores (weight) associated with the features. Depending on the data, plausible data analysis and preprocessing activities could be started to make it suitable for processing by the algorithm. For example, the textual dataset must be processed by apply typical NLP tasks (i.e., tokenization, stemming, stop-word removal, pos tagging, etc.), sentiment, and emotion analysis instead focuses on capturing the emotional aspect embedded in the word or sentence meaning. The EFCM algorithm works on data translated in a matrix form.

Each score given by the human experts and associated with a feature describes how that feature is relevant in the domain of interest, according to the expert viewpoint.

As shown in Fig. 2.1, the scores are processed (*Feature relevance (FR) Estimation*) to rescale them according to the appropriate range and evaluation metrics.

Acquired the input, the algorithm implements an iterative process: in each iteration, the EFCM is launched and, until the stability condition is not verified, the

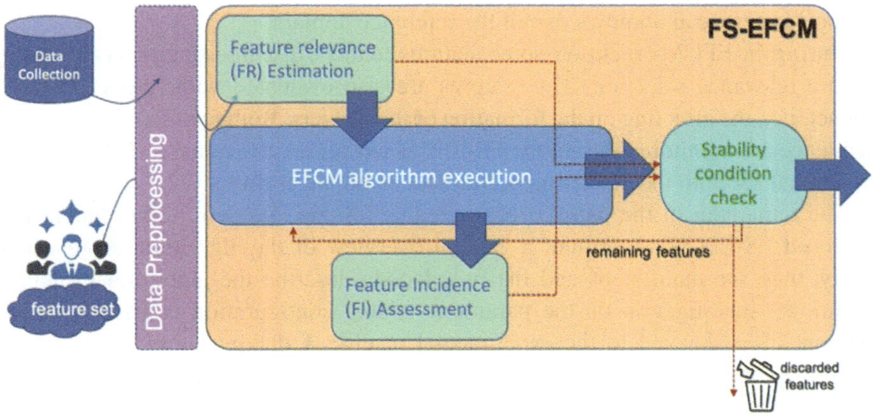

Fig. 2.1 Logical overview of the algorithm

whole algorithm is re-run by updating the parameter configuration. Precisely, the EFCM output is targeted at evaluating the *incidence* of each feature in the clustering formation (*Feature Incidence (FI) Assessment*).

The stability condition is strictly correlated to two important indices of the algorithm: the *feature relevance* FR and the *feature incidence* FI, which represents the importance of a feature from the human viewpoint and the incidence of the same feature from the data distribution in the clustering structure, respectively.

The algorithm stops when a condition of stability is reached, i.e., when all the features are strongly affecting the clustering formation.

Otherwise, when the stability condition is not satisfied, a further evaluation based on the two indices individuates the features candidate to be discarded. Once removed, the process is re-iterated con the remaining features.

2.2.1 EFCM Execution: Main Steps

The FS-EFCM algorithm can be described by the following steps:

1. *Feature relevance (FR) Estimation*: the collected score assigned by experts to the features s_h with $h = 1, \ldots, H$ are translated in a proper scale in the range [0, 1]. In general, scores assigned by experts could be defined in a scale correlated to the data domain. Thus, an index could be necessary to bound the feature score in the interval [0, 1]. For example, the score s_h could be fuzzified assigning a membership degree $\mu_{FR}(s_h)$ to a pre-defined fuzzy set (e.g., a sigma fuzzy set on a universe of a discourse given by an interval of the real line).
2. *EFCM algorithm execution* once given the data and feature scores, the EFCM algorithm is executed; in the first run, all the input features are used. The generated clusters are hyperspheres in the feature space.
3. *Feature incidence (FI) assessment:* the clusters generated by the EFCM are analyzed: the incidence of each feature in the clustering structure is calculated, by evaluating the *h*th feature component impact on each cluster prototype (measured as the distance of the feature components between the cluster prototype pairs). More formally, at the *t*th algorithm iteration, the weight value $w^{(t)}{}_h$ of hth feature component, is feature *incidence* value and it is calculated as follows:

$$w_h^{(t)} = \max_{\substack{i=1,2,\ldots,C^{(t)} \\ k=1,2,\ldots,C^{((t))} \\ i \neq k}} \left(\left| v_{ih}^{'(t)} - v_{kh}^{'(t)} \right| \right) \tag{2.1}$$

where v'_{ih} and v'_{kh} are the *h*th component values of all the cluster center pairs, evaluated at the *t*th iteration. The weight value $w^{(t)}{}_h$ assumes values between 0 and 1; the higher the value is, the more the feature affects the cluster formation. Similarly to the FR index, the incidence value is used to calculate the corresponding membership degree to a prefixed fuzzy set.

4. *Stability Condition Check:* The algorithm stops when the difference between the FI values of a feature in two successive iterations is below a prefixed threshold θ. Formally, the stability condition is given by:

$$\max_{h=1,\ldots,H}\left(\left|\mu_{FI}\left(w_h^{(t)}\right) - \mu_{FI}\left(w_h^{(t-1)}\right)\right|\right) \le \theta \tag{2.2}$$

   where $\mu_{FI}(w_h)$ is in general the membership degree to a fuzzy set defined on the feature w_h. If the condition holds, the current feature keeps being part of the feature set, otherwise the algorithm continues to the next step.
5. *Discard less significant features*: the remaining features are the candidate to be removed by the feature set since their contribution might be not significant in the clustering process. The selection of features candidate to be removed from the feature set is achieved by defining a new measure of significance μ_h of the *h*th feature for the clustering structure, applying the t-norm operator • as follows:

$$\mu_h = \mu_{FR}(s_h) \bullet \mu_{FI}(w_h) \quad h = 1, \ldots, H \tag{2.3}$$

   If μ_h is lower than a prefixed threshold δ, then the *h*th feature is unmeaningful, and then it is removed from the current feature set. Finally, the process returns to step 2, considering just the filtered features in the next iteration.

The pseudocode of the FS-EFCM is shown Listing 1.

Listing 1. Algorithm: *FS-EFCM*

Input: *FS={f₁,..f_H} feature set; N x H matrix MM*

```
Set δ, θ
Create the fuzzy set μFR(.) based on the feature relevance (FR) score values as-
signed by the experts
Repeat
   H := number of features    //initially all the features are considered
   Execute EFCM
      For h := 1 to H
         Calculate wh,
         Calculate μFI(wh)
         If μFR(sh) • μFI(wh) < δ
            Remove the h-th feature fh from FS
            H=H-1
Until the condition of stability is not reached
Return the partition matrix and the volume prototypes of the final C Clusters
```

2.2.2 Initial Parameter Setting

The parameter δ represents the threshold value below which a feature could be removed from the feature set. Its setting is crucial to controls the enhancement of the algorithm execution.

Thus, a preliminary step has been defined to set the proper value of the parameter δ. The rationale behind this step is to identify potential removable features, even though they were considered relevant by the experts.

Let n_R be the number of features considered relevant by the experts and n_D is the number of the candidate features to be discarded after the EFCM execution. If n_D is greater than 50% of n_R (i.e., many features that are relevant for the expert likely will be discarded), then the initial value of the parameter δ is re-set to a lower value.

Listing 2 shows the pseudocode of the algorithm *SelectDeltaThreshold* describing how the parameter δ is refined. Initially, n_R is calculated: it is the number of features considered very relevant by the experts, i.e., those with membership degree $\mu_{FR} \geq 0.7$ (lines 6–8). The value 0.7 is selected arbitrarily and it guarantee to select just the more meaningful features.

The relevant features candidate to be removed nD are stored in the array sig[] of nD size: lines 12–19 indeed describe how the array is filled. Line 13 shows the condition for discarding a feature: like in FS-EFCM Algorithm, μ_h of Eq. (2.3) must be lower than δ and, at the same time, checks if the feature is relevant ($\mu_{FR} \geq 0.7$). The feature significance value is added to the array sig[], if both conditions hold. Then, the array is ranked ascendingly, by feature significance values. The (non-zero) value in the intermediate (or the next non-zero intermediate) position of the array will be the new value of the threshold δ.

Let us notice that the value of the parameter δ must be lower than the initial value $\mu_{FR}(s_h)$, the membership degree assigned by the expert for the *h*th feature, and calculated for the FR fuzzy set. This guarantee that line 15 holds and further refinement of δ is applied in line 20.

This strategy ensures that more than half of the features relevant for the experts are not removed from FS-EFCM initially.

Listing 2. Algorithm: *SelectDeltaThreshold*

```
1.  Set δ, θ
2.  Create the fuzzy set μFR(.) based on the feature relevance (FR) score values as-
    signed by the experts
3.  H := number of features
4.  nR:= 0    // Initialize the number of relevant features
5.  nD:=0     // Initialize the number of relevant features potentially removable
6.  For h := 1 to H    // Calculate the number of relevant features
7.      If μFR(sh) ≥ 0.7
8.          nR:= nR +1
9.  Execute EFCM
10. For h := 1 to H
11.     Calculate wh
12.     Calculate μFI(wh)
13.     If μFR(sh) ● μFI(wh) < δ and μFR(sh) ≥ 0.7
14.         nD := nD +1
15.         sig[nD] := μFR(sh)● μFI(wh)
16.   Sort sig[] //the array sig[] is ascendingly sorted
17.   If nD / nR > 0.5
18.     med := nD /2   // the integer, intermediate value between 1 and nD
19.     If sig[med] > 0
20.       δ := sig[med]
21.     Else
22.       δ := the first not null value in sig[]
23.   Return δ
```

A further choice concerns the t-norm operator used in Line 13 of Listing 2 and defined in Step 5 of the FS-EFCM algorithm (Sect. 2.2.1). Among the families of t-norms defined in literature, the most widely used in application problems are the triangular norms, shown as follows:

- *minimum (Gödel)* t-norm $x \bullet y = \min(x, y)$
- *product (Goguen) t-norm* $x \bullet y = x \cdot y$
- *Lukasiewicz t-norm* $x \bullet y = \max(x + y - 1, 0)$

Depending on the selected t-norm, different fuzzy intersections are generated; in particular, the minimum t-norm is the most used in fuzzy controls, whereas the product t-norms produces a more drastic intersection than the minimum t-norm.

2.3 Experimental Results

2.3.1 Dataset

The experiments have been carried out on a Twitter dataset.

Twitter is one of the most preferred social networks, where people express their emotions, opinions about events, interests, items. It offers a comfortable platform for chatting, reading tweets, reacting accordingly by writing their comments through tweets, or sharing tweets written by others.

The dataset is composed of four hundred thousand public tweets posted by users from May 2018 to July 2018 in the cities of Washington, New York, and London [13]. All the tweets are in English and can have one or more hashtag. Tweets with the same hashtag are grouped in a unique document file: the hashtag allows the topic to be well-characterized. Files get as a name the hashtag that mainly appears in the collected tweets. Tweets without hashtag are discarded because they cannot be associated with a document.

The tweet collection has been preprocessed, applying elementary NLP tasks. Precisely, data have been stemmed, i.e., each word has been reduced to its inflectional form; then the stop words have been removed as well. Syntactic slangs and non-conventional dictionary terms are also discarded.

To apply the FS-EFCM algorithm, the data should be a term-documents matrix: each hashtag-based document is represented as a vector, and each cell contains a numeric value associated with a feature.

2.3.2 Feature Selection

In the text mining approaches, the feature set is usually composed of a collection of terms describing the data space peculiarities; often they are extracted by applying some metrics on the text (e.g., tf-idf-ranked top-k terms).

This experimentation aims at capturing emotions and human behavior from the tweet trends. From tweets analysis, it is possible to inspect people's sensations and feelings on local or world events; to investigate the main emotions and opinions that justify some behaviors.

Sentiments and opinions indeed are concealed in the sentences, typically associated with adjectives and verbs; then the intrinsic meaning of some textual expressions is not amenable to rigid linguistic patterns [2].

To this purpose, the feature set should be composed of a representative word set that describes the most relevant emotions.

A dataset[1] composed of sixty emotional categories (target classes) has been selected to cover a wide range of relevant emotional categories. A half of these categories describe *pleasant feelings* (they are labeled "open", "happy", "alive", "good", "love", "interest", "positive", "strong") whereas the remaining eight categories are related to *difficult/unpleasant feelings* ("angry", "depressed", "confused", "helpless", "indifferent", "afraid", "hurt", "sad"). Each emotional category is in turn composed of a set of words that "explain" the category; these words are generally synonyms or terms with similar meaning (e.g., "happy" and "glad"). The features set is comprehensively composed of 250 features, labeled with the words in the emotional categories.

---

[1] www.enchantedlearning.com.

2.3.3 FS-EFCM at Work

The FS-EFCM takes as input a matrix composed of 250 columns (term-features) and 5607 rows (documents extracted from a data stream of about four hundred and fifty thousand tweets, as stated above).

Each entry of the data point associated with a document vector is calculated by using the TF-IDF metrics.

For completeness, in Table 2.1, the name of all the features of the data points in our dataset.

To each data point (or document), one of the sixteen emotional categories is associated as the target class. This classification is obtained applying the method described in [6].

Initially, the experts assigned to each feature a relevance score in the range [1, 10]. In the initial configuration for running FS-EFCM, the threshold δ is set to 0.1, and the stop iteration/convergence threshold θ is set to 0.01.

Furthermore, we set the EFCM fuzzifier parameter to 2, the initial number of clusters to 50, the EFCM merging threshold to 0.01 and the EFCM stop iteration threshold to 0.001.

In the preprocessing phase, the *SelectDeltaThreshold* algorithm is executed to select the best value of δ.

After its execution, the significance of 135 features is below the prefixed threshold δ; just 20 of them are considered relevant for the expert (with a relevance index greater than or equal to 0.7). Then, the new value of δ is set to 0.05, and the *Select-DeltaThreshold* algorithm is re-run. 96 features, 9 of which are relevant to the expert, are removed. Table 2.2 summarizes these results.

Then, EFCM runs on the remaining 154 selected features. The algorithm stops after 8 cycles/iterations. In the last three cycles, the number of features decreases to 20, keeping this value until the last cycle. Table 2.3 shows the results obtained after each algorithm iteration.

After the eighth cycle the stop iteration value is less than the threshold θ and the algorithm stops.

In Table 2.4, the remaining twenty features and the measure of their significance are shown.

The final number of clusters is 16. The final document membership is calculated, assigning it to the cluster to which the document belongs with the highest membership degree.

Hence, cluster-class mapping is achieved by associating each cluster with the class to which most of the documents that have been assigned to the cluster belong to.

Table 2.5 shows for each emotional category, the number of documents in the class/category, the cluster label associated with the class, the number of documents assigned to this cluster, and the number of documents assigned to the cluster that belong to the class (well-classified documents).

Table 2.1 Features of the documents in our emotional dataset

Abomin	Absorb	Accept	Ach	Admir	Affect	Affection	Afflict	Afraid
Aggress	Agon	Alarm	Alien	Aliv	Alon	Amaz	And	Angri
Anguish	Anim	Annoy	Anxious	Appal	Asham	Attract	Bad	Bitter
Bless	Boil	Bold	Bore	Brave	Bright	Calm	Certain	Challenge
Cheer	Clever	Close	Cold	Comfort	Concern	Confid	Confus	Consider
Content	Courag	Coward	Cross	Crush	Curious	Dare	Deject	Delight
Depress	Depriv	Desol	Despair	Desper	Despic	Determin	Detest	Devot
Diminish	Disappoint	Discourag	Disgust	Disillus	Disinterest	Dismay	Dissatisfi	Distress
Distrust	Domin	Doubt	Drawn	Dull	Dynam	Eager	Earnest	Eas
Easi	Ecstat	Elat	Embarrass	Empti	Encourag	Energet	Engross	Enrag
Enthusiast	Excit	Fascin	Fatigu	Fear	Festiv	Forc	Fortun	Free
Frighten	Friski	Frustrat	Fume	Gay	Glad	Gleeful	Good	Great
Grief	Griev	Guilti	Happi	Hardi	Hate	Heartbroken	Helpless	Hesit
Hope	Hostil	Humili	Hurt	Import	Impuls	Incap	Incens	Indecis
Indiffer	Indign	Inferior	Inflam	Infuri	Injur	Inquisit	Insensit	Inspir
Insult	Intent	Interest	Intrigu	Irrit	Joyous	Jubil	Keen	Kind
Liber	Lifeless	Lone	Lost	Lousi	Love	Lucki	Menac	Merri
Miser	Misgiv	Mourn	Nervous	Neutral	Nonchal	Nosi	Offend	Offens
Open	Optimist	Overjoy	Pain	Panic	Paralyz	Passion	Pathet	Peac
Perplex	Pessimist	Play	Pleas	Posit	Powerless	Preoccupi	Provoc	Provok
Quak	Quiet	Re-enforc	Reassur	Rebelli	Recept	Reject	Relax	Reliabl
Repugn	Resent	Reserv	Restless	Sad	Satisfi	Scare	Secur	Sensit

(continued)

Table 2.1 (continued)

Seren	Shaki	Shi	Skeptic	Snoopi	Sore	Sorrow	Spirit	Stew
Strong	Stupefi	Sulki	Sunni	Sure	Surpris	Suspici	Sympathet	Sympathi
Tear	Tenaci	Tender	Tens	Terribl	Terrifi	Thank	Threaten	Thrill
Timid	Torment	Tortur	Touch	Toward	Tragic	Unbeliev	Uncertain	Understand
Uneasi	Unhappi	Uniqu	Unpleas	Unsur	Upset	Useless	Victim	Vulner
Wari	Warm	Weari	Woeful	Wonder	Worri	Wrong		

Table 2.2 Results obtained in the preprocessing phase

Number of features	Threshold δ	Features potentially removable	Relevant features potentially removable	New threshold δ	Feature removed	Features selected
250	0.1	135	20	0.05	96	154

Table 2.3 Results obtained launching FS-EFCM

Cycle	Number of features	Number of clusters	Feature removed	Features selected	Stop iteration value
1	154	20	62	92	0.42
2	92	18	37	55	0.23
3	55	16	22	33	0.15
4	33	16	9	24	0.09
5	24	16	3	21	0.05
6	21	16	1	20	0.02
7	20	16	0	20	0.013
8	20	16	0	20	0.005

Table 2.4 Significances of the final selected features

Feature name	Feature significance
Admir	0.09
Afraid	0.15
Anxious	0.18
Attract	0.11
Bad	0.16
Good	0.16
Great	0.13
Happi	0.15
Hurt	0.14
Love	0.10
Pain	0.21
Panic	0.12
Passion	0.15
Sad	0.13
Scare	0.08
Thank	0.22
Touch	0.14
Unhappi	0.15
Upset	0.13
Worri	0.08

Table 2.5 Number of documents assigned to the class and assigned to the correspondent cluster

Class	Documents belonging to the class	Cluster	Documents assigned to the cluster	Well classified documents
Happy	1380	C1	1369	1159
Open	366	C2	380	301
Confused	24	C3	31	13
Strong	55	C4	61	36
Interested	55	C5	49	35
Depressed	49	C6	41	26
Positive	421	C7	430	334
Hurt	377	C8	369	296
Good	428	C9	435	337
Alive	187	C10	182	141
Helpless	164	C11	166	128
Love	1186	C12	1175	1007
Angry	434	C13	426	352
Afraid	315	C14	322	257
Sad	161	C15	167	131
Indifferent	5	C16	4	3

About 81% of documents belonging to a class are assigned to the right cluster. The performance of FS-EFCM measured calculating the accuracy, precision, recall, and F1-score classification indices are shown in Table 2.6.

Now, let us consider just those documents whose membership degree to the cluster to which they are assigned is high or higher than a specific threshold; let us set that threshold to 0.6.

The number of documents whose membership degree to the assigned cluster is greater than 0.6 is 2047, which means it is about 37% of all the document collection.

Table 2.7 shows statistics similar to Table 2.5 but considering just the documents that strongly belong to the cluster they are assigned to.

Let us notice that the percent of documents belonging to a class and assigned to the correspondent cluster increase considerably (about 98%). Table 2.8 shows the classification performance for this document selection, evidencing notable improvements: all the metrics values are higher than 98%.

Table 2.6 Classification performance indices

Measure	Value (%)
Accuracy	77.02
Precision	77.88
Recall	77.21
F1-score	77.54

Table 2.7 Number of documents assigned to the class and assigned to the correspondent cluster considering the documents with membership degree to the assigned cluster greater than 0.6

Class	Documents belonging to the class	Cluster	Documents assigned to the cluster	Well classified documents
Happy	497	C1	494	487
Open	134	C2	133	128
Confused	10	C3	10	9
Strong	18	C4	20	17
Interested	24	C5	23	22
Depressed	21	C6	23	20
Positive	148	C7	149	143
Hurt	122	C8	119	116
Good	157	C9	158	153
Alive	75	C10	75	72
Helpless	56	C11	57	53
Love	450	C12	452	446
Angry	160	C13	162	157
Afraid	116	C14	114	110
Sad	57	C15	56	52
Indifferent	2	C16	2	2

Table 2.8 Classification performance indices calculated for the documents in Table 2.7

Measure	Value (%)
Accuracy	98.01
Precision	98.38
Recall	98.08
F1-score	98.23

These results highlight the effectiveness of the FS-EFCM, especially on supporting the classification of social data drawing relevant emotion-driven user behavior.

The experimentation reveals how the approach can generate accurate clusters, well-described by the features associated with each cluster. In fact, the final features are those that better represent the data distribution in the clusters and give a clear idea about the main sentiments and emotions expressed in the tweet trends analyzed.

2.4 Conclusion

This paper presents a semi-supervised clustering to classifying user generated. This paper presents a semi-supervised clustering to classifying user-generated content from the analysis of the main emotions expressed in the text.

The algorithm acquires from human experts some scores to assign a relevance degree to the features in the reference domain. Our experimentation focused on capturing emotions from tweet streams. For these reasons, the experts scored words expressing emotions or sentiments (such as "joy", "beautiful", etc.). The FS-EFCM algorithm reaches a trade-off between the feature relevance score provided by the experts and the feature impact on the cluster formation.

The performance of the algorithm revealed not just the effectiveness of the approach, but also its reliability in filtering very relevant features that clearly characterize the final clusters.

Future development of the algorithm aims at investigating deeply how the documents are associated with emotional categories. The idea is to consider not just the cluster to which the document belongs, with the highest membership degree, but also the other clusters to which the document can belong (with lower membership degree), to discover which sets of multiple emotional categories emerge from the document.

References

1. L. Alsumait, D. Barbara, C. Domeniconi, On-Line LDA: adaptive topic models for mining text streams with applications to topic detection and tracking, in *Conference: Proceedings of the 8th IEEE International Conference on Data Mining (ICDM 2008)*, Pisa, Italy (2008), 10 pp. https://doi.org/10.1109/ICDM.2008.140
2. L. Barbosa, J. Feng, Robust sentiment detection on twitter from based and noisy data, in *Proceedings of the 23rd International Conference on Computational Linguistics: Posters. COLING'10*, Beijing, China (Association for Computational Linguistics, Stroudsburg, PA, USA, 2010), pp. 36–44
3. J.C. Bezdek, *Pattern Recognition with Fuzzy Objective Function Algorithms* (Kluwer Academic Publishers, Norwell, MA, USA, 1981). https://doi.org/10.1007/978-1-4757-0450-1
4. J.C. Bezdek, R. Ehrlich, W. Full, The fuzzy C-means clustering algorithm. Comput. Geosci. **10**(2–3), 191–203 (1984)
5. E. Cambria, Affective computing and sentiment analysis. IEEE Intell. Syst. **31**(2), 102–107 (2016)
6. D. Cavaliere, S. Senatore, V. Loia, Context-aware profiling of concepts from a semantic topological space. Knowl. Based Syst. **130**, 102–115 (2017)
7. G. Chandrashekar, F. Sahin, A survey on feature selection methods. Comput. Electr. Eng. **40**(1), 16–28 (2014). https://doi.org/10.1016/j.compeleceng.2013.11.024. ISSN 0045-7906
8. K. Dave, S. Lawrence, D.M. Pennock, Mining the peanut gallery: opinion extraction and semantic classification of product reviews, in *Proceedings of the 12th International Conference on World Wide Web, WWW '03* (ACM, New York, NY, USA, 2003), pp. 519–528. https://doi.org/10.1145/775152.775226
9. I. Guyon, A. Elisseeff, A., An introduction to variable and feature selection. J. Mach. Learn. Res. **3**(2003), 1157–1182 (2003)

10. U. Kaymak, M. Setnes, Fuzzy clustering with volume prototype and adaptive cluster merging. IEEE Trans. Fuzzy Syst. **10**(6), 705–712 (2002)
11. B. Liu, Sentiment analysis and opinion mining. Synth. Lect. Human Lang. Technol. **5**(1), 168 (2012). https://doi.org/10.2200/S00416ED1V01Y201204HLT016
12. C.D. Manning, P. Raghavan, H. Schutze, *Introduction to Information Retrieval* (Cambridge University Press, New York, 2008), 506 pp. ISBN 978-0521865715
13. F. Martino Di, S. Senatore, S. Sessa, A lightweight clustering-based approach to discover different emotional shades from social message streams. Int. J. Intell. Syst. **1**, 19 (2019). https://doi.org/10.1002/int.22105
14. D.W. Otter, J.R. Medina, J.K. Kalita, A survey of the usages of deep learning for natural language processing. IEEE Trans. Neural Netw. Learn. Syst. (2020). https://doi.org/10.1109/TNNLS.2020.2979670
15. S. Poria, A. Gelbukh, A. Hussain, N. Howard, D. Das, S. Bandyopadhyay, Enhanced senticnet with affective labels for concept-based opinion mining. IEEE Intell. Syst. **28**(2), 31–38 (2013). https://doi.org/10.1109/MIS.2013
16. A. Rehman, J. Kashif, H.A. Babri, S. Mehreen, Relative discrimination criterion—a novel feature ranking method for text data. Expert Syst. Appl. (Elsevier) **42**, 3670–3681 (2015)
17. G. Salton, C. Buckley, Term-weighting approaches in automatic text retrieval. Inf. Process. Manag. **24**(5), 513–523 (1988). https://doi.org/10.1016/0306-4573(88)90021-0
18. L. Stanchev, Semantic document clustering using a similarity graph, in *2016 IEEE Tenth International Conference on Semantic Computing (ICSC)* (2016), pp. 1–8. https://doi.org/10.1109/ICSC.2016.8
19. C. Strapparava, R. Mihalcea, Learning to identify emotions in text, in *Proceedings of the 2008 ACM Symposium on Applied Computing, SAC'08* (ACM, New York, NY, USA, 2008), pp. 1556–1560. https://doi.org/10.1145/1363686.1364052
20. S. Sun, C. Luo, J. Chen, A review of natural language processing techniques for opinion mining systems. Inf. Fusion **36**, 10–25 (2017). https://doi.org/10.1016/j.inffus.2016.10.004
21. T. Wei, Y. Lu, H. Chang, Q. Zhou, X. Bao, A semantic approach for text clustering using WordNet and lexical chains. Expert Syst. Appl. **42**(4), 2264–2275 (2015). https://doi.org/10.1016/j.eswa.2014.10.023
22. G. Miller, C. Fellbaum, WordNet. An electronic lexical database. Cambridge, MA: MIT Press; (1998). 423 pp, ISBN: 978-0262061971.

Chapter 3
AI in (and for) Games

Kostas Karpouzis and George A. Tsatiris

Abstract This chapter outlines the relation between artificial intelligence (AI)/machine learning (ML) algorithms and digital games. This relation is two-fold: on one hand, AI/ML researchers can generate large, in-the-wild datasets of human affective activity, player behaviour (i.e. actions within the game world), commercial behaviour, interaction with graphical user interface elements or messaging with other players, while games can utilise intelligent algorithms to automate testing of game levels, generate content, develop intelligent and responsive non-player characters (NPCs) or predict and respond to player behaviour across a wide variety of player cultures. In this work, we discuss some of the most common and widely accepted uses of AI/ML in games and how intelligent systems can benefit from those, elaborating on estimating player experience based on expressivity and performance, and on generating proper and interesting content for a language learning game.

Keywords Machine learning · Artificial intelligence · Games · Procedural content generation · Affective computing · Player behaviour · Computational culture

3.1 Introduction

Digital games have enjoyed a huge wave of popularity among the research community in the past years. An important reason for this is the fact that games combine the characteristics and requirements of performance/narrative media [11] with high-maintenance software and hardware requirements for storage, CPU performance,

K. Karpouzis (✉)
Department of Communication, Media and Culture Panteion University of Social and Political Sciences, Athens, Greece
e-mail: kkarpou@cs.ntua.gr

G. A. Tsatiris
Artificial Intelligence and Learning Systems Laboratory,
National Technical University of Athens, Athens, Greece
e-mail: gtsatiris@image.ntua.gr

G. A. Tsihrintzis et al. (eds.), *Advances in Machine Learning/Deep Learning-based Technologies*, Learning and Analytics in Intelligent Systems 23,
https://doi.org/10.1007/978-3-030-76794-5_3

network communication [51] and security [32]. Especially in the field of computer hardware, digital games have been a driving force for the industry to produce newer, more effective and less power-demanding hardware for PCs and game consoles [45], pushing their capabilities further and further.

Another important fact that makes digital games extremely popular as a research medium has to do with the relative ease to find participants for games research studies: Pew Research [46] mentions that "43% of U.S. adults say they often or sometimes play video games on a computer, TV, game console or portable device", with puzzle and strategy games constituting the most popular genres among those who often or sometimes play video games. As a result, researchers working with games, either in the core of their work or as a means to attract users and record their behaviour or preferences, can quickly put together large corpora of data (e.g. [24] or [55] for a database which captures player expressivity associated with player behaviour or [18] for a 3D dataset describing tennis-related actions in video and 3D skeleton form).

Among the research areas that embraced digital games as a platform of choice, perhaps the most celebrated one has been the combination of Artificial Intelligence (AI) and Machine Learning (ML), mostly because of the popularity of AI/ML algorithms which competed against and eventually beat human world champions in Chess [9] and more complex board games such as Go [58, 59]. Games are a fitting medium to train and test AI/ML algorithms because of the relatively small search space in which to look for and identify the best possible move and, mostly, for the completeness and robustness of the definition of the game world in terms of variables, rules and relations. Conversely, game design and development has been putting to use AI/ML algorithms to create content automatically or in a user-guided manner, to estimate and adapt player experience [19, 73] or predict player behaviour [12]. In this chapter, we will start with identifying possible sources of data to be used to train and test AI/ML algorithms; in the following, we will discuss the areas of conversation between intelligent algorithms and digital games, providing examples of identification of player behaviour and prediction of player experience, and elaborate on game content generation for serious games in education.

3.2 Game Content and Databases

Perhaps the most important component for the interplay between AI/ML, along with the actual algorithms and their context or use case, is the selection and role of data or content to be used or generated. In game design and development terminology, *content* refers to a wide variety of concepts, data and types of media within a game world. An interesting distinction is that besides content being included in the game as part of its design or interactive functionality, it can be generated by the game during game play [1], usually as a response of the game world to the choices and actions performed by the player, or it can be created by the players themselves [26]. The latter case is usually referred to as "user" (UGC) or "player-generated content" and covers personalisation of the appearance of the player character (PC)

Fig. 3.1 Example of User Generated Content, where players choose the appearance of their avatar

in the game world (Fig. 3.1), choices that relate to game actions [69], commercial behaviour (e.g. buying digital goods or aesthetic elements using real money or in-game digital currency) or interaction with other players, usually using text chat, voice communication or even sign language [10] capabilities.

An interesting trait of player-generated content is that, as research shows [8], it is less controlled by social filters and inhibition; this effectively means that players who participate and function within the safe sandbox of a digital game world express themselves more vividly [15], using a wider range of spontaneous emotions [25] and microexpressions [20] than usual affect- and emotion-related interactions [23], thus offering richer input for AI/ML algorithms and catering for their deployment "in-the-wild" [14, 41]. An example of spontaneous user-generated content was presented in the Platformer Experience Dataset (or PED[1]) [24]. This is a multimodal dataset which contains videos of 58 participants playing IMB [66], an open-source clone of the popular "Super Mario Bros" platformer game (Fig. 3.2), recordings of the game screen synchronised with the videos, logs of player actions with timestamps and a self-reported assessment of fun, interest and player experience in two forms, ratings and ranks (see [75] for an interesting discussion on *ranking* different choices or preferences, instead of *rating* each of them).

[1] The dataset can be downloaded from https://ped.institutedigitalgames.com/.

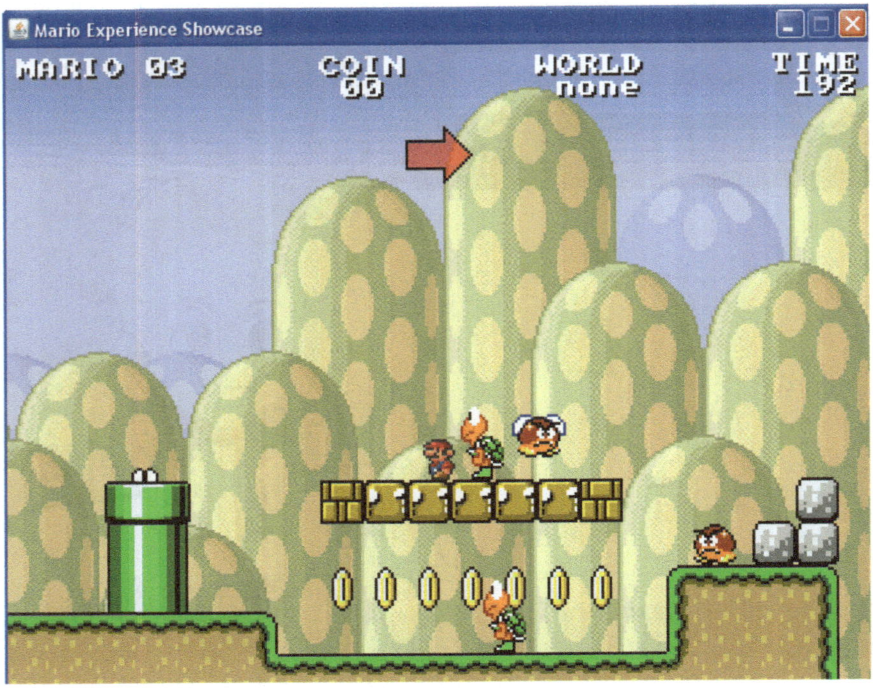

Fig. 3.2 Generated game level in IMB, used during the recording of the PED dataset

The approach of recording player expressivity along with player behaviour (actions in the game world) and player experience allows for a number of possible uses of AI/ML algorithms, either on recognition/classification or on generation of game content. For example, Asteriadis et al. combined head and body expressivity from players (Fig. 3.3) with in-game actions to cluster them in different groups [3], taking also game performance and demographics into account [2]. An interesting observation of this work was that players often used microexpressions or microgestures in conjunction with their actions and the relevant movement of their character; for instance, they would nod in sync with *jump* actions or tilt their head to the direction of movement in response to avoiding an enemy. Clustering players with respect to expressivity and performance also identified interesting patterns: for instance, expert players were either very expressive, in the sense that they were immersed in the game action and mimicked their character's movement with body movements, or almost inanimate, indicating a high level of concentration. Overall, this work identified the need to combine player behaviour with expressive analysis [25] so as to produce meaningful and dependable results regarding player engagement. In the same framework, Pedersen et al. [39] produced levels for IMB which were predicted to be fun and engaging for each particular player, based on their affective and behavioural input while playing. This concept combines the sensing and player experience prediction work with modelling the aspects of the game level that make it *hard*, *fun* or *irritating*

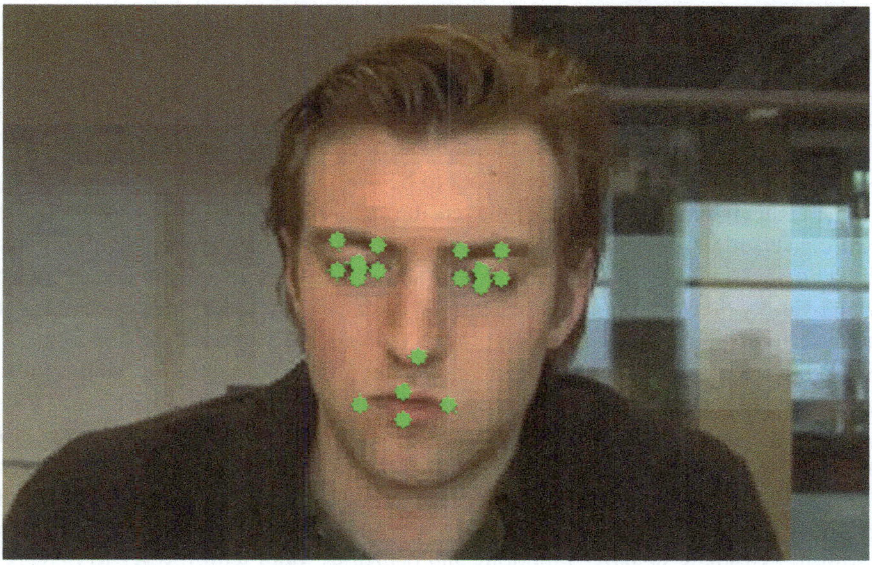

Fig. 3.3 Video frame showing player expressivity and detected facial features

for each player: in the context of platform games, such as SMB or IMB, these factors include the number of gaps in a given level, the gap size, the number and placement of enemies and the positioning of rewards and power-ups. This work showed that by mapping player experience modelling with difficulty modelling, content generation algorithms can create individual game levels with a high degree of probability to be interesting and engaging.

Besides affective or audiovisual data of people playing games, the most relevant source of data comes from player behaviour (actions within the game world). The amount of data produced by players during gameplay differs with each genre, with turn-based or strategy games producing a few samples per minute and action games or Real-time Strategy (RTS) games providing hundreds of individual player actions per minute (APMs): in "Starcraft", one of the most popular RTS games ever produced, top players typically record around 400 APMs in the preparation phase and close to 800 APMs during battle. As a result, accumulating large corpora of data and using them to train different AI/ML architectures has been a very popular use case for games and ML researchers. For example, Ravari et al. [44] utilise the datasets presented in [49, 63] to predict the winner of each match, identifying relevant and important time-dependent (e.g. player actions, such as *build* or *attack*) and time-independent features, such as buildable areas in each game map or height of specific areas; to achieve this, they employ Gradient Boosting Regression Trees (GBRT) [16] and Random Forest (RF) [6] implementations in Scikit-learn, an open-source Python package. In the same context, Lin et al. produced a very large dataset of more than 65.000 games, which also includes visual information, besides player behaviour

[30]; the sheer amount of data included here (1535 million frames, 496 million player actions) illustrates the relevance of player behaviour data to Big Data algorithms and processes (cf. [4] on churn prediction, i.e. when players quit playing a particular game, or [74] on how player behaviour data are processed in the game industry to promote player experience and spending behaviour).

A more contemporary source of data to be used with AI/ML algorithms comes from players interacting with other players during or before gameplay, in the form of text conversation (chat) or using voice. As mentioned earlier, gameplay eliminates most of the social inhibitions in players and allows for richer interaction and a wider variety of extreme emotions. Murnion et al. [35] utilise commercial sentiment analysis tools, such as Twinword Sentiment Analysis and Microsoft Azure Cognitive Services, to process player behaviour and game logs from an online multi-player game called World of Tanks (WoT); the authors are looking for positive vs. negative interactions and specific abusive behaviours, such as derogatory insults or racist attacks. An important aspect of this work is that it uses easily accessible services to retrieve, decrypt and process the data, making the study easy to replicate and extend. The context of the study is also very interesting, since cyberbullying can result to extremely negative emotions and decisions, both with respect to players' real lives, as well as their sense of attrition from the game: different studies have shown that more than 50% of players have either quit or considered quitting a game because of cyberbullying behaviours [17]. A similar approach regarding game data was used in [36], where authors used topic modelling and statistical analysis to analyse interactions between viewers (or spectators) of matches in Dota,[2] a real-time multiplayer battle game. Their work identified patterns similar to those of football match spectators, especially in the case of intra-audience effects, despite the fact that audience commentary in eSports does not reach (and, hence, influence) the players directly.

3.3 Intelligent Game Content Generation and Selection

Game content generation has been a very active field in which AI/ML showcase their potential when it comes to generating data, mainly since the term *content* can refer to most audio, visual and narrative concepts in a game. Besides visual appearance, such as environment aesthetics or 2D/3D models of characters and environments, game content may refer to audio or aural media (e.g. the soundtrack of the game or specific audio effects used in response to game events, such as firing a weapon or player death), graphical user interfaces (GUI) [43], where interactive elements are used by the game to convey information (e.g. that another player or enemy is nearby) or by players to select game options and engage in game behaviour, or even the game narrative itself. The latter case [21, 37, 50]) is extremely interesting, since

[2] A richer dataset that includes 50,000 matches, game data, player skill ratings and chat can be found at https://www.kaggle.com/devinanzelmo/dota-2-matches.

it caters for the generation of different stories and games, based on an initial plan or narrative by the game designer. The success of narrative generation techniques in Role-Playing Games (RPGs), which are mainly popular in Japan, shows that this content generation option has the potential to create longer-lasting gameplay and keep players motivated, serving the needs of both researchers and the industry [60]. A special case of narrative generation includes NPC planning and behaviour [53], with content generation algorithms choosing and executing non-player character actions and dialogues based on virtual personalities (e.g. a *sidekick* or an enemy) or player behaviour (e.g. respond to the player ransacking a hut which belongs to a friendly villager in an RPG).

Intelligent content generation techniques can also be categorised with respect to the level of automation they require or provide. Some of those techniques enable designers to be involved in the process, either by initiating major changes in the way generated content is evolved [29] or by allowing players to adapt the content generated in the game [57]. Fully automated techniques [56], usually referred to as Procedural Content Generation (PCG), have been extremely popular with researchers, since they require little or no input, besides fine-tuning the respective algorithm parameters, but they also enjoy success in commercial games: in the 1980s, dungeon games such as *Akalabeth* and *Rogue* were among the first to use automatic (but sometimes random) content generation, while *Elite* (1985), a 3D space exploration game, used content generation to author 8 galaxies with 256 solar systems each and 1–12 planets in each solar system, all within 32 Kb of code. Between *Diablo* (1995) and the recent years, PCG was mostly constrained to RPGs and dungeon layouts, with automatic content creation being revived by Minecraft (2011), developed by Mojang and now owned by Microsoft. More recent games include *Left for Dead* (2008) (instantiating game objects such as trees, monsters or treasures), *S.T.A.L.K.E.R.: The Shadow of Chernobyl* (2007) (dynamic systems create unscripted NPC behaviour), *Apophenia* (2008) (generation of puzzles and plots), and mainly *No Man's Sky* (2016) with a procedurally generated deterministic open world universe and planets with unique flora and fauna (Fig. 3.4), and various sentient alien species.

Automatic PCG is usually matched with a respective generation algorithm and relevant constraints that match the game design or simply make sense. Vocabulary- or grammar-based generation algorithms are usually deployed to create game worlds, mazes or dungeons [22] and plants, whose recursive shape lends well to the way generation works in cellular automata [33] or L-systems [67]. More recently, Müller et al. used L-systems to [34] to create 3D buildings with different sizes, number of rooms or floors. This is an interesting example, since the way the L-system is initially authored may reflect specific rules or constraints (e.g. to create a variety of structurally sound buildings in [72] or even be used to generate complete cities [64]). Given the amount of polygonal geometry and texture images each building may require in a 3D environment, an automatic PCG algorithm that is able to create huge worlds with very little overhead in memory or CPU usage can replace the need to author the necessary objects beforehand. This is also the case with algorithms based on Artifical

Fig. 3.4 Planet and environment generation in No Man's Sky

Life approaches: BIOME[3] is a programmable cellular automata simulator that allows users to develop simple "SimCity-like" grids, simulating phenomena such as forest fires, disease epidemics or animal migration patterns (cf. Fig. 3.5).

When it comes to *sequences* of data to be generated, for instance in procedural music generation or when planning behaviours for NPCs, Hidden Markov Models (HMMs) are a usual choice. Snodgrass et al. [61] generated sequences of game levels for different platform games, comparing their performance in each one, while Plans et al. [42] combined generation with player experience to author the music score for a game. More recently, Long Short-Term Memory (LSTM) networks have been used to generate game levels, e.g. in a Super Mario Bros. clone [62]. LSTMs seem to have taken over the PCGML (Procedural Content Generation based on Machine Learning) experiments, thanks to the readily available implementations in Python and C#, as well as their recurrent nature that caters for generation of diverse and (theoretically) infinite content [48]. For instance, Savery and Weinberg [52] used LSTMs to synthesise musical scores based on image and video analysis, and Botoni et al. [5] to create NPCs with more depth in terms of dialogue and style.

3.3.1 Generating Content for a Language Education Game

Generation of appropriate content for serious/educational games is an extremely important concept since it can make all the difference between adoption and retainment of the game, thus increasing the possibility to achieve its learning objectives,

[3] Download BIOME from http://www.spore.com/comm/prototypes.

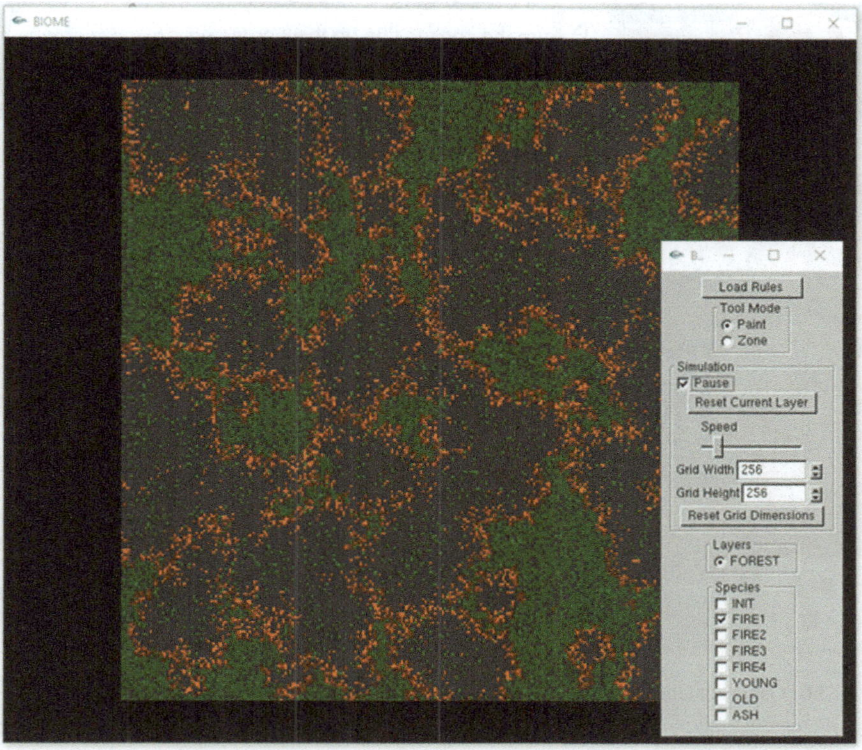

Fig. 3.5 Generated forest fire in BIOME

and attrition [71]. In the iRead project,[4] we are creating a serious game and supporting applications for entry-level language learning of English, English as a Foreign Language (EFL), German, Spanish and Greek [38]. The core software applications developed in the project are a reader application,[5] which highlights parts of the words contained in the text, given specific criteria, and a serious game,[6] which consists of a series of gamified activities utilising words and sentences. The foundation of these applications and the software infrastructure that provides access to the content consists of language models for each language, including for children with dyslexia; following the definition of extensive phonological and syntactic models for these languages, the linguists in the project worked with teachers to define the learning objectives for each of the target age groups, as well as the sequence in which each language feature should be taught [31]. The sequencing of these features, including which prerequisites should be taught and mastered by the students before moving on to more advanced features, was encoded in a tree-like hierarchical graph; essentially,

[4] iRead project. https://iread-project.eu/.

[5] Amigo reader application, https://iread-project.eu/amigo-reader/.

[6] Navigo game. https://iread-project.eu/game/.

Fig. 3.6 Game content generation for a Navigo mini-game

this graph encapsulates both the language model (i.e. the features that make up each word or sentence, at least at the given language level) and the teaching model, represented by the selection of necessary features for each school year and the succession in which they should be taught. When a new student registers with the iRead system, this graph is instantiated as a user profile, with different values of mastery for each feature, depending on the student's age.

This is where the adaptive content generation component [68] in iRead kicks in, first by utilising the mastery levels for each feature to select proper content from the project resource engine (dictionaries and texts) and then by updating the student's model based on their performance in each language game they play; when the mastery level for a given feature surpasses a selected threshold (75%), subsequent features in the model hierarchy become available to play with, provided that all prerequisites for them have been met. In the context of iRead, the game content consists of selecting a particular game activity; a language feature to work with; and a set of words or a sentence that corresponds to that feature (e.g. a particular letter, phoneme or a sequence of phonemes) (Fig. 3.6).

The content selection process starts (cf. Fig. 3.7 for an outline of the process) with the given student model, i.e. the mastery level for each open feature; then it selects the content for each session by filtering the available resources with a set of rules defined by project researchers after productive consultation sessions with the teachers

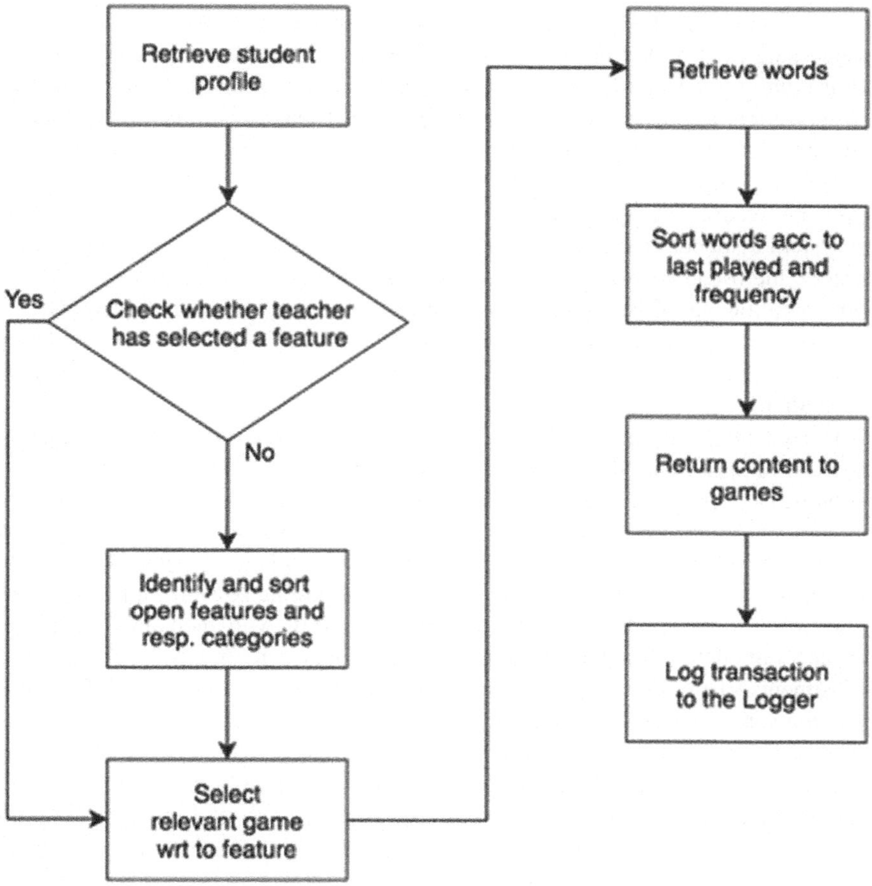

Fig. 3.7 Flowchart of actions to generate content in iRead

collaborating with them. These rules were first defined in verbal form, in order to promote the teaching objectives of the games, with each of them corresponding to a particular pedagogical rationale. For instance, when multiple features are open (available to play with), the Adaptation component sorts them by taking into account how many times each of the features has been used in earlier games and how well the student has previously performed when presented with that feature. The reasoning here is that students should start from an easier feature and should not be playing a feature they have not done well with recently, thus fostering motivation and efficacy. Other rules attempt to reinforce learning by combining the feature mastery level achieved in previous games with the number of gameplay sessions since that feature was last used, and by reopening a fully mastered feature after ten games have been played since it was last used. The assumption is that the student has fully mastered that feature, but they must repeat it once in a while, to showcase their progress and long-term mastery. Finally, a number of feature selection rules deal with students not progressing as expected or not having truly mastered the language features which correspond to their age level: if a feature has been practised twice and the feature mastery is not improving, the mastery level for that feature and its prerequisites is reduced, so that both can be revisited in future sessions. This content selection strategy treats the assumption that students in a given age group have already mastered certain language features, by allowing them to go back to required knowledge, if there is no system evidence that it has been acquired. In addition to selecting proper word content, the iRead adaptivity system utilises a rule-based strategy to select specific game activities to utilise those words: if a feature has not been previously used in a game for the particular student, then the selected game should promote accuracy in using that language characteristic, before moving on to games which stimulate automaticity.

The second part of the adaptation component in iRead has to do with re-evaluating the value of mastery for the language feature used in a game. During the consultation sessions, teachers mentioned a number of requirements for this process: changes in mastery values should not be abrupt, especially when students make an occasional error in one of the activities; besides this, they should help students demonstrate complete mastery of a feature within a handful of gaming sessions, allowing them to move on to more advanced and interesting features. A mathematical definition that accommodates these requirements, while leaving room for experimentation and adjustments of the process, is that of Exponential Moving Average (EMA) [40]: essentially, this takes into account previous attempts at a particular feature (previous game sessions) but gives more weight to recent attempts. The number of previous attempts to consider may be defined by each implementation; in iRead, we implemented the complete definition, but chose to consider only the previous value of mastery, when calculating the next one. This averaging process allows students to show complete mastery in just three games, since each newly opened feature is initialised with a value of 5 and after three perfect games reaches the maximum value of 10. In addition, in case the student makes one or more errors during game play, the respective mastery value may be reduced by 1 at maximum. This mechanic, along with the rules which prioritise features given the recent gameplay attempts, allows students to practise different language content, without being stuck with difficult features.

On-going evaluation [7, 47] has shown that the automated content selection and the profile re-evaluation processes are quite close to what teachers expect and provide suitable and interesting content for the students. Even though the re-evaluation mechanic allows unlocking subsequent features quite easily, there have been reports that students are being given the same features to play with during numerous successive game sessions. However, after going through the system logs of gameplay results and mastery evaluations, we reached the conclusion that this reflects the design of the respective language model, which imposed a large number of prerequisites to be unlocked before moving on to more complex features. This effectively illustrates the interplay between the different iRead components: the features which describe each language model, the graph of prerequisites which describes the sequencing embedded in the learning process, the mastery levels for each feature which reflect student performance, and the adaptation and re-evaluation rules described above, which prioritise content to implement teaching objectives.

The large-scale evaluation phase in schools across Europe is already underway, with more than 2000 students taking part. Even though it has been disrupted by the pandemic and schools closing down, we expect gameplay logs to keep coming in from students playing the games at home. Processing these logs will allow us to revisit specific parts of the adaptation component, primarily the content selection rules and the mastery re-evaluation implementation.

3.4 Conclusions

The interplay between AI/ML algorithms and digital games has been in the forefront of scientific news and research outlets for the past few years. The main reason for this is that it fosters adaptive player experiences [76], which promote and even maximise fun and engagement. Besides this, AI/ML can be used to select and generate diverse and related game content, even from sources of Open and Big Data (cf. [70] for a Monopoly clone populated with Open Data [65] to teach Big Data rankings and associations in the context of a primary school geography course or [13] for an approach that uses Open Data in a card game for environmental education), making games more relevant to everyday life. In this chapter, we outlined some of the more prominent approaches which combine AI/ML with game design concepts and player behaviour to provide information about player experience and generate content that's predicted to maximise engagement. We also described an approach to estimate player experience and engagement based on behaviour and affective expressivity during gameplay and an intelligent algorithm that generates language content for a serious game, based on player performance and learning/teaching objectives. As users provide more and richer input to AI/ML algorithms through explicit choices and gameplay, it is expected that this interplay will become even more meaningful and will be integrated in more applications, expanding into education, inclusion [54] and gamification ([27, 28]).

Acknowledgements This work has been partly funded by the iRead project which has received funding from the European Union's Horizon 2020 Research and Innovation programme under Grant Agreement No. 731724.

References

1. A. Amato, Procedural content generation in the game industry, in *Game Dynamics* (Springer, 2017), pp. 15–22
2. S. Asteriadis, K. Karpouzis, N. Shaker, G.N. Yannakakis, Does your profile say it all? using demographics to predict expressive head movement during gameplay (2012)

3. S. Asteriadis, K. Karpouzis, N. Shaker, G.N. Yannakakis, Towards detecting clusters of players using visual and gameplay behavioral cues. Procedia Comput. Sci. **15**, 140–147 (2012)
4. P. Bertens, A. Guitart, Áf. Periáñez, Games and big data: a scalable multi-dimensional churn prediction model, in *2017 IEEE Conference on Computational Intelligence and Games (CIG)* (IEEE, 2017), pp. 33–36
5. B. Bottoni, Y. Moolenaar, A. Hevia, T. Anchor, K.A. Benko, R. Knauf, K.P. Jantke, A.J. Gonzalez, A.S. Wu, Character depth and sentence diversification in automated narrative generation, in *FLAIRS conference* (2020), pp. 21–26
6. L. Breiman, Random forests. Mach. Learn. **45**(1), 5–32 (2001)
7. L. Bunting, Y.H. Segerstad, W. Barendregt, Swedish teachers' views on the use of personalised learning technologies for teaching children reading in the english classroom. Int. J. Child-Comput. Int. **27**, 100236 (2021)
8. P. Cairns, A.L Cox, M. Day, H. Martin, T. Perryman, Who but not where: the effect of social play on immersion in digital games. Int. J. Hum. Comput. Stud **71**(11), 1069–1077 (2013)
9. M. Campbell, A.J. Hoane Jr, F.-H. Hsu, Deep blue. Artif. Intell. **134**(1–2), 57–83 (2002)
10. G. Caridakis, S. Asteriadis, K. Karpouzis, Non-manual cues in automatic sign language recognition. Pers. Ubiquit. Comput. **18**(1), 37–46 (2014)
11. E. Carstensdottir, E. Kleinman, M.S. El-Nasr, Player interaction in narrative games: structure and narrative progression mechanics, in *Proceedings of the 14th International Conference on the Foundations of Digital Games* (2019), pp. 1–9
12. D. Charles, B.U. Cowley, Behavlet analytics for player profiling and churn prediction, in *International Conference on Human-Computer Interaction* (Springer, 2020), pp. 631–643
13. D. Chiotaki, K. Karpouzis, Open and cultural data games for learning. *arXiv* preprint arXiv:2004.07521 (2020)
14. R. Cowie, C. Cox, J.-C. Martin, A. Batliner, D. Heylen, K. Karpouzis, Issues in data labelling, in *Emotion-oriented systems* (Springer, 2011), pp. 213–241
15. K. Durning, *Jackson College of Graduate Studies, and Jackson College of Graduate Studies* (Department of Psychology. Gaming Relationship to Social Psychology and Microexpressions, University of Central Oklahoma, 2016)
16. J.H. Friedman, Stochastic gradient boosting. Comput. Stat. Data Anal. **38**(4), 367–378 (2002)
17. M. Fryling, J.L. Cotler, J. Rivituso, L. Mathews, S. Pratico, Cyberbullying or normal game play? impact of age, gender, and experience on cyberbullying in multi-player online gaming environments: perceptions from one gaming forum. J. Inf. Syst. Appl. Res. **8**(1), 4 (2015)
18. S. Gourgari, G. Goudelis, K. Karpouzis, S. Kollias, Thetis: three dimensional tennis shots a human action dataset, in *Proceedings of the IEEE Conference on Computer Vision and Pattern Recognition Workshops* (2013), pp. 676–681
19. C. Guckelsberger, C. Salge, J. Gow, P. Cairns, Predicting player experience without the player an exploratory study, in *Proceedings of the Annual Symposium on Computer-Human Interaction in Play* (2017), pp. 305–315
20. S.H. Hemenover, N.D. Bowman, Video games, emotion, and emotion regulation: expanding the scope. Ann. Int. Commun. Assoc. **42**(2), 125–143 (2018)
21. S. Imabuchi, T. Ogata, A story generation system based on propp theory: as a mechanism in an integrated narrative generation system, in *International Conference on NLP* (Springer, 2012), pp. 312–321
22. L. Johnson, G.N. Yannakakis, J. Togelius, Cellular automata for real-time generation of infinite cave levels, in *Proceedings of the 2010 Workshop on Procedural Content Generation in Games* (2010), pp. 1–4
23. K. Karpouzis, G.N. Yannakakis, *Emotion in Games* (Springer, 2016)
24. K. Karpouzis, G.N. Yannakakis, N. Shaker, S. Asteriadis, The platformer experience dataset, in *2015 International Conference on Affective Computing and Intelligent Interaction (ACII)* (IEEE, 2015), pp. 712–718
25. I. Kotsia, S. Zafeiriou, G. Goudelis, I. Patras, K. Karpouzis, Multimodal sensing in affective gaming, in *Emotion in Games* (Springer, 2016), pp. 59–84

26. G. Lastowka, User-generated content and virtual worlds. Vand. J. Ent. & Tech. L. **10**, 893 (2007)
27. N.Z. Legaki, K. Karpouzis, V. Assimakopoulos, Using gamification to teach forecasting in a business school setting, in *GamiFIN* (2019), pp. 13–24
28. N.-Z. Legaki, N. Xi, J. Hamari, K. Karpouzis, V. Assimakopoulos. The effect of challenge-based gamification on learning: An experiment in the context of statistics education. Int. J. Hum. Comput. Stud. 102496 (2020)
29. A. Liapis, G. Smith, N. Shaker, Mixed-initiative content creation, in *Procedural Content Generation in Games* (Springer, 2016), pp. 195–214
30. Z. Lin, J. Gehring, V. Khalidov, G. Synnaeve, Stardata: a starcraft ai research dataset. *arXiv* preprint arXiv:1708.02139 (2017)
31. M. Mavrikis, A. Vasalou, L. Benton, C. Raftopoulou, A. Symvonis, K. Karpouzis, D. Wilkins, Towards evidence-informed design principles for adaptive reading games, in *Extended Abstracts of the 2019 CHI Conference on Human Factors in Computing Systems* (2019), pp. 1–4
32. T. Min, W. Cai, A security case study for blockchain games, in *2019 IEEE Games, Entertainment, Media Conference (GEM)* (IEEE, 2019), pp. 1–8
33. S. Miremadi, B. Lennartson, K. Akesson, A BDD-based approach for modeling plant and supervisor by extended finite automata. IEEE Trans. Control Syst. Technol. **20**(6), 1421–1435 (2011)
34. P. Müller, P. Wonka, S. Haegler, A. Ulmer, L. Van Gool. Procedural modeling of buildings, in *ACM SIGGRAPH 2006 Papers* (2006), pp. 614–623
35. S. Murnion, W.J. Buchanan, A. Smales, G. Russell, Machine learning and semantic analysis of in-game chat for cyberbullying. Comput. Secur. **76**, 197–213 (2018)
36. I. Musabirov, D. Bulygin, P. Okopny, K. Konstantinova, Between an arena and a sports bar: online chats of esports spectators. *arXiv preprint*arXiv:1801.02862 (2018)
37. T. Ogata, Building conceptual dictionaries for an integrated narrative generation system. J. Robot. Netw. Artif. Life **1**(4), 270–284 (2015)
38. D. Panagiotopoylos, A. Symvonis, iread: infrastructure and integrated tools for personalized learning of reading skill. Inf. Intell. Syst. Appl. **1**(1), 44–46 (2020)
39. C. Pedersen, J. Togelius, G.N. Yannakakis, Modeling player experience for content creation. IEEE Trans. Comput. Intell. AI Games **2**(1), 54–67 (2010)
40. R. Pelánek, J. Řihák, Experimental analysis of mastery learning criteria, in *Proceedings of the 25th Conference on User Modeling, Adaptation and Personalization* (2017), pp. 156–163
41. B. Perron. A cognitive psychological approach to gameplay emotions (2005)
42. D. Plans, D. Morelli, Experience-driven procedural music generation for games. IEEE Trans. Comput. Intell. AI Games **4**(3), 192–198 (2012)
43. R. Popp, D. Raneburger, H. Kaindl, Tool support for automated multi-device GUI generation from discourse-based communication models, in *Proceedings of the 5th ACM SIGCHI Symposium on Engineering Interactive Computing Systems* (2013), pp. 145–150
44. Y.N. Ravari, S. Bakkes, P. Spronck, Starcraft winner prediction, in *Twelfth Artificial Intelligence and Interactive Digital Entertainment Conference* (2016)
45. J.P. Research (2017) *eSports is a Driving Force Behind PC Gaming Hardware Sales Growth*. Accessed 3 September 2020. https://www.globenewswire.com/news-release/2017/07/11/1042645/0/en/JPR-eSports-is-a-Driving-Force-Behind-PC-Gaming-Hardware-Sales-Growth.html
46. P. Research, *5 facts about Americans and video games* (2018). https://www.pewresearch.org/fact-tank/2018/09/17/5-facts-about-americans-and-video-games/. Accessed 3 September 2020
47. A. Révész, M. Vasalou, A. Florea, R. Gilabert, L. Bunting, Y.H. Segerstad, I. Mihu, C. Parry, L. Benton, The effects of textual enhancement on development in l2 derivational morphology: a multi-site longitudinal study (2020)
48. S. Risi, J. Togelius, Procedural content generation: from automatically generating game levels to increasing generality in machine learning. *arXiv preprint*arXiv:1911.13071 (2019)

49. G. Robertson, I. Watson, An improved dataset and extraction process for starcraft ai, in *The Twenty-Seventh International Flairs Conference* (Citeseer, 2014)
50. J. Robertson, R. Michael Young, Automated gameplay generation from declarative world representations, in *AIIDE*, pp. 72–78 (2015)
51. S. S. Sabet, S. Schmidt, S. Zadtootaghaj, C. Griwodz, S. Moller, Towards the impact of gamers strategy and user inputs on the delay sensitivity of cloud games, in *2020 Twelfth International Conference on Quality of Multimedia Experience (QoMEX)* (IEEE, 2020), pp. 1–3
52. R. Savery, G. Weinberg, Shimon the robot film composer and deep score: an LSTM for generation of film scores based on visual analysis. *arXiv preprint*arXiv:2011.07953 (2020)
53. A. Savidis, There is more to PCG than meets the eye: NPC AI, dynamic camera, PVS and lightmaps. *arXiv preprint*arXiv:1808.00328 (2018)
54. A. Schmölz, K. Karpouzis, D. Pfeiffer, P. Koulouris, Doing social inclusion: aiming to conquer crisis through game-based dialogues and games
55. N. Shaker, S. Asteriadis, G.N. Yannakakis, K. Karpouzis, A game-based corpus for analysing the interplay between game context and player experience, in *International Conference on Affective Computing and Intelligent Interaction* (Springer, 2011), pp. 547–556
56. N. Shaker, J. Togelius, M.J. Nelson, *Procedural Content Generation in Games: A Textbook and an Overview of Current Research* (Springer, 2016)
57. N. Shaker, G.N. Yannakakis, J. Togelius, Towards player-driven procedural content generation, in *Proceedings of the 9th Conference on Computing Frontiers, CF '12* (New York, NY, USA, 2012), pp. 237–240. Association for Computing Machinery
58. D. Silver, A. Huang, C.J. Maddison, A. Guez, L. Sifre, G. Van Den Driessche, J. Schrittwieser, I. Antonoglou, V. Panneershelvam, M. Lanctot et al., Mastering the game of go with deep neural networks and tree search. Nature **529**(7587), 484–489 (2016)
59. D. Silver, T. Hubert, J. Schrittwieser, I. Antonoglou, M. Lai, A. Guez, M. Lanctot, L. Sifre, D. Kumaran, T. Graepel et al., A general reinforcement learning algorithm that masters chess, shogi, and go through self-play. Science **362**(6419), 1140–1144 (2018)
60. G. Smith, E. Gan, A. Othenin-Girard, J. Whitehead, PCG-based game design: enabling new play experiences through procedural content generation, in *Proceedings of the 2nd International Workshop on Procedural Content Generation in Games* (2011), pp. 1–4
61. S. Snodgrass, S. Ontañón, Learning to generate video game maps using Markov models. IEEE Trans. Comput. Intell. AI Games **9**(4), 410–422 (2016)
62. A. Summerville, S. Snodgrass, M. Guzdial, C. Holmgård, A.K. Hoover, A. Isaksen, A. Nealen, J. Togelius, Procedural content generation via machine learning (pcgml). IEEE Trans. Games **10**(3), 257–270 (2018)
63. G. Synnaeve, P. Bessiere, A dataset for starcraft ai & an example of armies clustering (2012)
64. J.O. Talton, Y. Lou, S. Lesser, J. Duke, R. Měch, V. Koltun, Metropolis procedural modeling. ACM Trans. Graph. (TOG) **30**(2), 1–14 (2011)
65. S. Theocharis, G.A. Tsihrintzis, Ontology development to support the open public data-the greek case, in *IISA 2014, The 5th International Conference on Information, Intelligence, Systems and Applications* (IEEE, 2014), pp. 385–390
66. J. Togelius, S. Karakovskiy, R. Baumgarten, The 2009 mario ai competition, in *IEEE Congress on Evolutionary Computation* (IEEE, 2010), pp. 1–8
67. J. Togelius, N. Shaker, J. Dormans, Grammars and l-systems with applications to vegetation and levels, in *Procedural Content Generation in Games* (Springer, 2016), pp. 73–98
68. G. Tsatiris, K. Karpouzis, Developing for personalised learning: the long road from educational objectives to development and feedback, in *ACM Interaction Design and Children (IDC) Conference 2020, Workshop on Technology-Mediated Personalized Learning for Younger Learners: Concepts, Methods and Practice* (2020)
69. G.A. Tsihrintzis, D.N. Sotiropoulos, L.C. Jain, Machine learning paradigms: advances in data analytics, in *Machine Learning Paradigms* (Springer, 2019), pp. 1–4
70. I. Vargianniti, K. Karpouzis, Using big and open data to generate content for an educational game to increase student performance and interest. Big Data Cogn. Comput. **4**(4), 30 (2020)

71. M. Virvou, G. Katsionis, K. Manos, Combining software games with education: evaluation of its educational effectiveness. J. Educ. Technol. Soc. **8**(2), 54–65 (2005)
72. E. Whiting, J. Ochsendorf, F. Durand, Procedural modeling of structurally-sound masonry buildings, in *ACM SIGGRAPH Asia 2009 Papers* (2009), pp. 1–9
73. J. Wiemeyer, L. Nacke, C. Moser et al., Player experience, in *Serious Games* (Springer, 2016), pp. 243–271
74. M. Willson, T. Leaver, Zynga's farmville, social games, and the ethics of big data mining. Commun. Res. Pract. **1**(2), 147–158 (2015)
75. G.N. Yannakakis, R.Cowie, C. Busso, The ordinal nature of emotions, in *2017 Seventh International Conference on Affective Computing and Intelligent Interaction (ACII)* (IEEE, 2017), pp. 248–255
76. G.N. Yannakakis, K. Isbister, A. Paiva, K. Karpouzis. Guest editorial: emotion in games. IEEE Trans. Affect. Comput. **5**(1), 1–2 (2014)

Part II
Machine Learning/Deep Learning in Education

Chapter 4
Computer-Human Mutual Training in a Virtual Laboratory Environment

Vasilis Zafeiropoulos and Dimitris Kalles

Abstract Science universities face the constant challenge of training their students in making appropriate use of their laboratories while avoiding accidents or equipment damages. Such a task becomes even harder for universities offering distance education, as their students visit less often their lab facilities and may have limited opportunities to become familiar with the respective instruments and equipment. For this purpose, the Hellenic Open University, which offers a long-standing distance education program in natural sciences, has been developing Onlabs, an interactive 3D virtual lab resembling its on-site biology laboratory for its students to train before they actually conduct live experiments. Recent versions of Onlabs contain, among others, an Instruction Mode, in which the human user is being guided by the computer to conduct a particular simulated experiment, and an Evaluation Mode, in which the computer evaluates the performance of the human user with respect to the completion of an experiment, contributing further to an effective learning process. Hence, in order for the performance assessment to be accurate, two separate machine learning techniques, a genetic algorithm and back-propagation on an artificial neural network, have been used.

Keywords Genetic algorithms · Artificial neural networks · Virtual labs · Distance education · Laboratory training

4.1 Introduction

Onlabs is an interactive 3D virtual lab simulating the on-site biology laboratory of Hellenic Open University and is used for the training of undergraduate and postgraduate students in biology. The students make use of it at home, before they come to the physical lab.

V. Zafeiropoulos · D. Kalles (✉)
Hellenic Open University, Patras, Greece
e-mail: kalles@eap.gr

G. A. Tsihrintzis et al. (eds.), *Advances in Machine Learning/Deep Learning-based Technologies*, Learning and Analytics in Intelligent Systems 23,
https://doi.org/10.1007/978-3-030-76794-5_4

Fig. 4.1 A screenshot of Onlabs (version 2.1.2)

Onlabs consists in fact of a modern adventure game, incorporating state-of-the-art 3D graphics and having the interaction done through the keyboard and the mouse while the user is expected to follow particular tasks.

4.1.1 Purpose and Development of the Virtual Lab

Science universities face the challenge of training their students in making successful use of their laboratories and conducting experiments. Yet there is always the risk of accidents and equipment damages which restricts the thorough use of the instruments. At the same time, science students at distance universities are confronted with additional barriers, such as making rare use of the on-site laboratory facilities, thereby being inefficiently trained.

For those reason, Hellenic Open University (HOU), which offers distance education in Natural Sciences as well as other disciplines, has been developing a 3D virtual biology lab, Onlabs, for its students to use before they make use of the on-site lab. Onlabs offers students the opportunity to experiment with the virtual lab equipment and learn by trial and error.

Onlabs has been being developed under the high-end 3D game-engines Hive3D[1] (2012–2015) and Unity (2016-now). A screenshot of its latest stable version, 2.1.2, released in November 2019, is shown in Fig. 4.1. Currently, a beta edition of version 3.0 is also available while its stable edition is expected to be released in 2021.

[1] Released by Eyelead, a computer game company in Athens, Greece.

4.1.2 Different Playing Modes

Onlabs's latest version 2.1.2 contains two distinct simulated experimental procedures and three different modes of playing. The procedures are those of *microscoping of a test specimen*, in which the user is supposed to set up the microscope, create a test specimen and microscope it with all objective lenses of the photonic microscope; and the *preparation of 500 ml of 10X TBE water solution*, for which the user weights 17.4 gr of Boric Acid and 54 gr of Trizma Base powders on the electronic scale, dissolves them in water with the magnetic stirrer and adds extra water as well as EDTA pH 8.0 to the produced solution. The playing modes are those of *instruction*, where the user is instructed by voice and text in completing the selected procedure, each time being allowed to perform only the suggested move; *evaluation*, where the user can make any move they want within the selected procedure and receives an evaluation for their performance; and *experimentation*, where the user may make any action they want (provided it is included in the game) and handles all the instruments and equipment from both procedures without being evaluated. Onlabs version 2.1.2 can be downloaded and run from the project's website.[2]

Onlabs version 3.0 contains two new experimental procedures which respectively supersede the procedures in version 2.1.2; *microscoping*, which mainly includes the microscoping of a test specimen of the previous versions, and *electrophoresis*, whose first part is the preparation of 500 ml of 10X TBE water solution.

Apart from the afore-mentioned student versions, we have also released Onlabs Machine Learning version 1.0, which, as its name suggests, is used for machine learning training. Onlabs ML 1.0 concerns only the microscoping procedure and instead of *experimentation mode* of playing, it contains a new one, that of *computer training mode*. The latter in fact contains two different kinds of training; *training the rater*, consisting of several play sessions by the human user followed by the calibration of the assessment mechanism upon which the human user is evaluated; and *training the bot*, in which a bot, or, in computer games vocabulary, a Non-Playable Character (NPC), performs random actions within the selected experimental procedure and is being simultaneously trained. Rater training uses either an interactive Genetic Algorithm (GA), meaning a GA adjusted to supervised learning, or back-propagation on an Artificial Neural Network (ANN), whereas bot training is done with the use of Reinforcement Learning (RL).

In this paper, after giving a short description of Onlabs's conceptual design and embedded assessment mechanism, we present the two machine learning techniques used in rater training.

---

[2] http://onlabs.eap.gr/.

4.1.3 Evaluation

First version of Onlabs, v0.1, released in 2013, was tested and evaluated by HOU's undergraduate biology students; for the evaluation, System Usability Questionnaire (SUS) for systems engineering, developed by John Brooke in 1996 was used [1]. The evaluation results suggested that Onlabs was easy to use and the students who later on used the on-site biology lab showed an improved performance [2, 3]. Thereafter, versions 0.2, 0.2.1 and 0.2.3 were evaluated by undergraduate students at other universities as well as high school science teachers with the use of an adjusted version of the afore-mentioned questionnaire. The evaluation produced similar results as the one of version 0.1.

Version 2.1, released in 2017, was more thoroughly evaluated. Concerning the experimental procedure of microscoping of a test specimen, two educational methods were used and examined in comparison to each other; the routine on-site lab tutorial method and the use of Onlabs along with a running Skype session. The students who took part produced some particular learning results which were later evaluated with Pre-Tests and Post-Tests. Pre-Tests scores confirmed that students who used Onlabs broadened their baseline knowledge whereas Post-Tests ones demonstrated explicitly that the performance of students having used Onlabs was better that the performance of those having not used it [4, 5].

4.2 Background and Related Work

Onlabs, being an adventure game, is suitable as a test-bed for artificial agents' training; in fact, already since the 2000s, adventure games have been proposed for that purpose [6, 7]. Various machine learning techniques have been used within adventure game applications specifically designed for that purpose. One of them is *Sophie's Kitchen*, developed by Thomaz and Breazeal [8, 9]. There, an NPC representing a robot called Sophie has the goal to combine the necessary ingredients to make a cake and bake it in the oven; the machine learning method that is used for Sophie's training is reinforcement learning.

Adventure games and text-based ones in particular, have recently been proposed again as potential machine learning test-beds. In text-based adventure games, various machine learning techniques have in fact been applied for the NPCs' training, like deep reinforcement learning [10, 11] and ANNs [12, 13].

Despite ANNs not being popular for computer games because of their black-box nature [14], there have been various studies and applications of them in gaming, for the classification of opponents [15], the exploration of arcade game levels [16] and the control of the fight or flight responses of NPCs [14]. In the last two decades, GAs have also been proposed for computer games, for the purpose of adjusting the NPCs' behavior [15, 17–19].

GAs simulate biological evolution; they therefore had no direct connection to supervised learning. Nevertheless, interactive GAs, which consist of a special kind of GAs that are suitable for supervised learning, have been proposed in several researches [20–23]. The GA we use in Onlabs is of this kind.

4.3 Architecture of the Virtual Laboratory

As mentioned in the introduction, Onlabs's latest stable version 2.1.2 has three different modes; instruction, evaluation and computer training. We successively demonstrate all of them. However, we first need to cast light on our virtual lab's architecture that those modes are based upon as well as the machine learning algorithms that are used in the computer training mode. In this section, we demonstrate Onlabs's conceptual design and its embedded assessment mechanism.

4.3.1 Conceptual Design

The virtual environment of Onlabs is discrete, meaning that the various interactable objects and instrument parts are distinct from each other and can be handled by the user through specific actions. It is also deterministic, meaning that an object at a given state, after a new action is made upon this or another object, produces a uniquely defined new state, and by inference, the overall state of the environment is also uniquely defined. The classification of Onlabs' virtual environment as discrete and deterministic has been done according to the criteria set by Russell and Norvig [24].

Various *entities* exist within the virtual environment of Onlabs. Those can be either *characters*, that is, the *Ego Character*, controlled by the human user, and the NPC, controlled by the computer with Artificial Intelligence, or *objects*. Entities are represented by classes design-wise as well as development-wise. Classes are abstract and each one of them represents one or more kindred entities. For instance, the various bottles (water bottle, EDTA bottle) are instances of the *bottle* class. Moreover, a class can be a specialization of another, generic one, that is, have various particular characteristics along with the generic ones; for instance, the *bottle* class is a specialization of the *vessel* class. For simplicity reasons, each entity will be mentioned with their class name, with the exception of when we deal with two or more kindred entities, where their names will be used.

The human user through *Ego* and the computer through the NPC are allowed to perform particular actions on the interactable objects. Those are *Pick Up* (picking up an object to the inventory), *Press* (pressing a button, switch or trigger), *Release* (releasing a pressed button, switch or trigger), *Rotate* (rotating a knob), *Withdraw* (withdrawing an object from its position, e.g. a plug from a socket), *Spin* (spinning an object that can be spun, e.g. a volumetric cylinder with water in order to dissolve powder in it) and *Combine With* (combining an object with another one).

A class may have qualitative features, whose acceptable values are alphanumeric or quantitative ones, whose acceptable values are numeric. For instance, the qualitative *state* feature of *AC switch* class (representing the AC switch of the microscope) gets the values 'ON' and 'OFF' whereas the quantitative *position* feature of *aperture knob* (representing the aperture knob of the microscope) gets integer values between 0 and 40. Those values can be changed directly, by actions that the human user through *Ego* or the computer through the NPC performs on the respective objects (for example, by pressing the *AC switch* of the microscope, its *state* feature changes from 'OFF' to 'ON' or 'ON' to 'OFF', depending on its previous value); or indirectly, by actions performed on other objects whose features' values are changed and simultaneously cause the change of the features' values in question (for example, by rotating the microscope's *coarse focus knob*, its *angle* feature is being changed along with the *height* feature of the *stage base*, resulting to the latter's moving up or down).

The feature-based architecture of entities follows Poole and Mackworth's guidelines [25]. Our past works about the conceptual design of Onlabs [2, 3] can also be accessed on our website.

4.3.2 State-Transition Diagrams

For the purpose of depicting the changes in the various alphanumeric feature values of Onlabs's entities, we use the notation of State-Transition Diagrams (STDs). In general, STDs are used for representing state changes in both time-critical and time-independent systems [26]. They are also called State Diagrams (SDs) but we chose the STD variation in order to lay significance on transitions (and therefore, the actions that cause them) in Onlabs. Furthermore, STDs comprise of a useful tool for the domain experts (in our case, biologists), who most of the time have no knowledge in computing science, to model their experiments with the help of the knowledge engineers and the programmers.

In our STD design, the states are depicted by circles and the transitions by arrows; the state that a transition arrow starts from is called the *initial state* of this particular transition while the state it leads to is called the *final state*. Figures 4.2 and 4.3

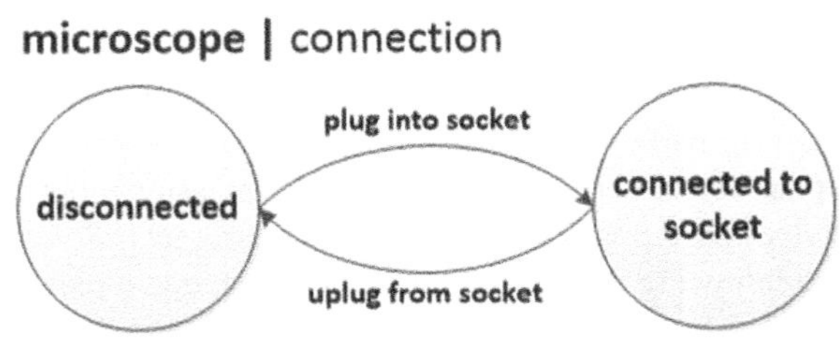

Fig. 4.2 STD of *microscope*'s *connection* feature

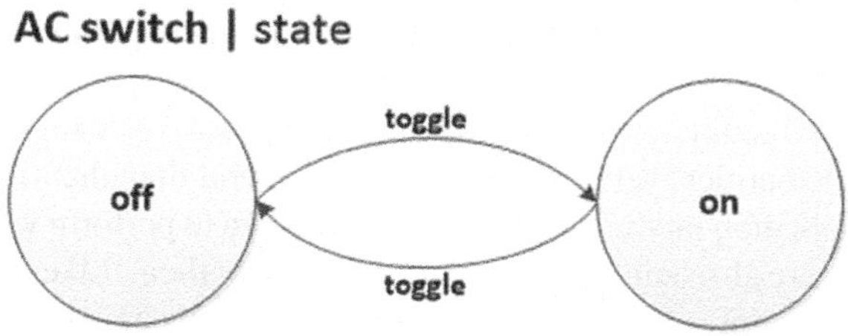

Fig. 4.3 STD of *AC switch*'s *state* feature

Fig. 4.4 STD of *slide*'s *condition* feature

represent the STDs for *microscope*'s *connect* and *AC switch*'s *state* (not to be confused with the 'state' notion of the STDs), respectively.

A transition from a state of an entity's feature to another state can be triggered either by an action the user performs on this particular entity or by an action on another entity whose feature state change sends a signal which ultimately reaches this one. For example, in the case of *microscope AC switch*'s *state* (Fig. 4.2), both *toggle* transitions are triggered by the *Press* action on the AC switch; on the other hand, in the case of *microscope*'s *connection* (Fig. 4.3), the *plug into socket* transition is triggered by using the microscope's plug with a socket while the *unplug from socket* one is triggered by the *Withdraw* action on the plug, provided the latter has been connected to a socket before.

The afore-mentioned STDs were binary, transiting from one state to the other. However, not all our STDs in Onlabs's design are like that. For example, the STD for *slide*'s *condition* feature (depicted in Fig. 4.4), contains four different states and a one-way path of transitions connecting them.

4.3.3 High Level Design

Up to now, Onlabs's design has concerned the various lab objects and their attributes and manipulation; in other words, it has been *low level*. In order for the design to comply with a hierarchical agent architecture [25], a high-level description of more general tasks performed by the ego character must also be introduced.

The high-level design has two forms, each one corresponding to the respective play mode: the *instruction* and the *evaluation* ones.

Under the instruction mode, the user is each time allowed to perform only the action than they are being guided to. The guidance takes places in both text and audio and it specifically follows the steps of the experiment on focus. The key to this design is a State Machine comprising of a particular STD for the *Ego Character*'s *state* feature, which is similar to the low-level ones referring to the various inanimate entities, yet more abstract and general than them.

Under the evaluation mode, the user has the freedom to perform whichever action they like but they are also being evaluated for each one of them. Like in the instruction mode but in a much more complicated way, this particular design is based on a general evaluation algorithm which takes the steps of the experiment on focus into account.

4.3.4 State Machine

For the purpose of illustrating the State Machine in the instruction mode, the microscoping procedure will be examined. The latter consists of setting the microscope on and testing its various parts as well as creating a test specimen using a paper scrap and water and microscoping it with all objective lenses (4X, 10X, 40X and 100X). According to HOU's biology laboratory manual, the particular steps that *Ego* needs to follow in order to successfully carry out the afore-mentioned set of tasks are shown in Table 4.1.

Those 25 steps are actually encoded into the respective transitions in the STD of *Ego*'s *state* feature while the states are named accordingly (Fig. 4.5).

As shown in Fig. 4.5, The Ego STD is non-branched and unidirectional, meaning that only a particular action is permitted each time; for example, in the beginning, only the joint use of *microscope plug* with *socket* is allowed, and this produces the 'connect microscope to socket' transition and leads from the 'beginning' state to the 'microscope connected to socket' one. Nevertheless, there are also steps that allow for more than one action; for example, at the 'entered microscoping mode' state, 4 different entities are allowed for handling (specifically *Rotate*), those of *coarse focus knob*, *fine focus knob*, *stage knob*, *specimen holder knob*, *condenser knob*, *left ocular* and *right ocular*, and their correct adjustment leads to accomplishment of 'focus with lens 4X' transition and consequently, the 'focused with lens 4X' state.

Obviously, as soon as a transition is made, there is no way reversing them. Such restriction is obviously necessary for the instruction mode that we are dealing with.

Although the Ego's STD at first glance looks like the objects' ones, it has two crucial differences from them. The first difference is the one we have already mentioned, that while an object's STD transition corresponds to a single action, an Ego transition may also refer to a set of actions. The other difference is that, unlike in the case of the various object features, the naming of the states of the *Ego*'s STD is conventional and has only to do with the steps to be followed by the Ego Character. For example, the 'microscope connected to socket' state is named exclusively after the 'connect microscope to socket' transition that leads to it, and that, obviously, does not mean that in all succeeding states the microscope is not connected to the socket; on the contrary, it remains connected to the socket until a new transition causes its disconnection, which, in our case, is the last one in the diagram.

Table 4.1 The necessary steps for the successful completion of the microscoping procedure in the instruction mode (steps with asterisks and ampersands are omitted in the evaluation)

1	Connect the microscope into the socket
2	Turn the microscope light on
3	Set light intensity to 3 4ths of maximum
4	Set iris fully open
5	Lift the condenser lens to the top position
6	Set lens 4X active
7	Test coarse focus knob
8	Test fine focus knob
9	Test stage knob
10	Test specimen holder knob
11	Prepare test specimen *
12	Put test specimen on stage
13	Enter microscoping mode
14	Focus with lens 4X
15	Set lens 10X active *
16	Focus with lens 10X
17	Set lens 40X active *
18	Focus with lens 40X
19	Set lens 100X active *
20	Focus with lens 100X
21	Exit microscoping mode #
22	Collect specimen #
23	Close iris #
24	Turn off light #
25	Disconnect microscope from socket #

4.3.5 Individual Scores

As we have already mentioned, the *individual scores* we have designed for Onlabs are used for the user's assessment in *evaluation mode*, that is, for evaluating the human user's performance with respect to a specific experimental procedure, and for the training of the rater in *training mode*, that is, calibrating both kinds of assessment mechanism used. Those are the weights in either the weighted average of the individual scores or the ANN; the weights of the former are calibrated with the use of an interactive GA while the calibration of the weights of the latter is done with Back-Propagation algorithm.

Individual scores are supposed to range from 0 to 1 so that their weighted average ranges from 0 to 1, too. Obviously, 0 and 1 are the worst and the best values respectively an individual score can get.

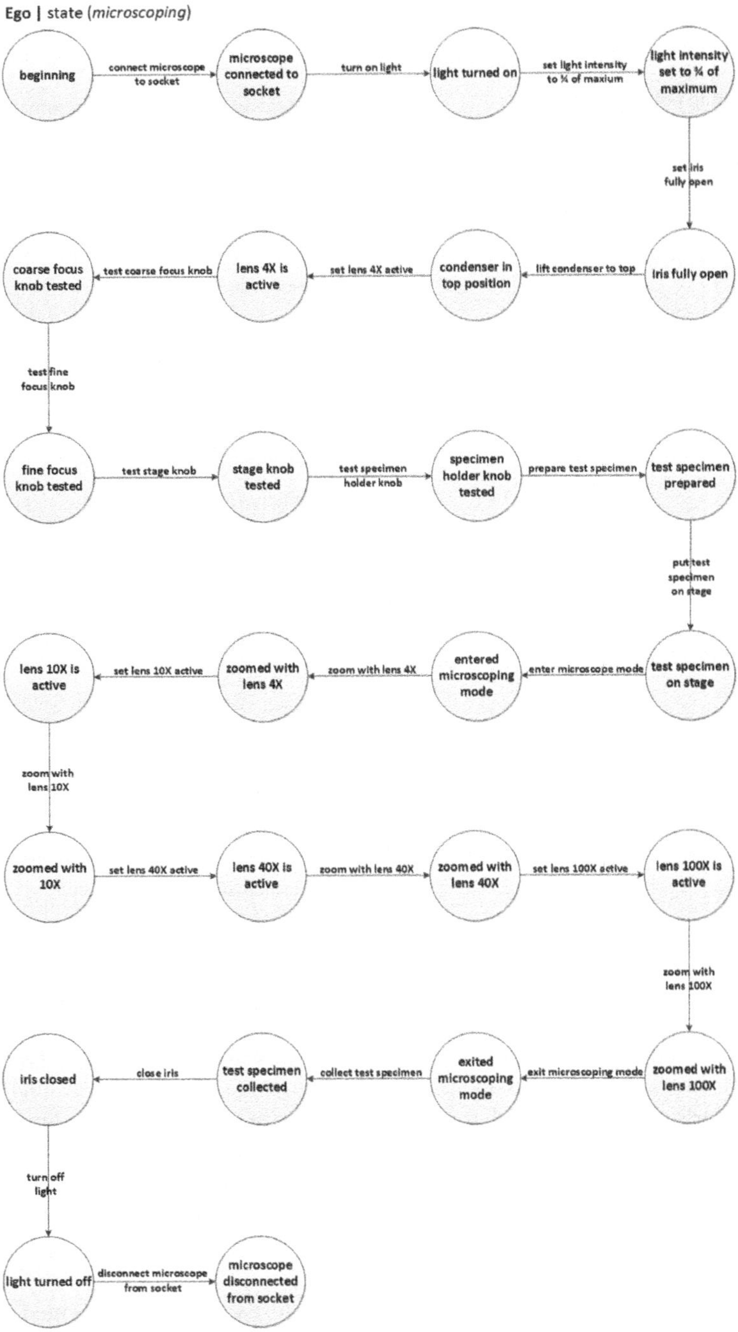

Fig. 4.5 The *Ego*'s state feature STD for the microscoping procedure

Table 4.2 The necessary steps for the successful completion of microscoping procedure

1	Connect the microscope into the socket *microscope|connection status* ← 'connected to socket'
2	Turn the microscope light on *AC switch|state* ← 'ON'
3	Set light intensity to 2 3rds of maximum *light intensity knob|position* ← ¾ • MaxPosition
4	Set iris fully open *aperture knob|position* ← MaxPosition
5	Lift the condenser lens to the top position *condenser|height* ← MaxHeight
6	Set lens 4X active *revolving nosepiece|active focus* ← 4
7	Test coarse focus knob *coarse focus knob|tested* ← 'yes'
8	Test fine focus knob *fine focus knob|tested* ← 'yes'
9	Test stage knob *stage knob|tested* ← 'yes'
10	Test specimen holder knob *test specimen knob|tested* ← 'yes'
11	Put test specimen on stage *stage|attached specimen* ← 'yes'
12	Enter microscope mode *Ego|mode* ← 'microscoping'
13	Focus with lens 4X *Microscoping view|clearness* [*revolving nosepiece|active focus* = 4]← 1
14	Focus with lens 10X *Microscoping view|clearness* [*revolving nosepiece|active focus* = 10]← 1
15	Focus with lens 40X *Microscoping view|clearness* [*revolving nosepiece|active focus* = 40]← 1
16	Focus with lens 100X *Microscoping view|clearness* [*revolving nosepiece|active focus* = 100]← 1

We will now describe our individual score architecture through the microscoping procedure. The necessary steps for the successful completion of the procedure are those listed in Table 4.2, which contains the steps in Table 4.1 excluding the ones marked either with an asterisk (*) or an ampersand (#); the former are considered rather redundant in terms of design in the evaluation mode while the latter are dealt with by a separate subprocess, that of resetting of instruments, which will be discussed later on.

We denote individual scores by x_k, $k = 1 \ldots n$, where n is the number of features to be altered (16 in our example).

Most values in Table 4.2 are quantitative whereas some of them are qualitative. In order to calculate the individual scores for all features in the list, we need to carry out the processes of *quantification, normalization* and *composite evaluation*. Below we describe those processes through various examples from the table.

4.3.6 Quantization

Step 1 shown in Table 4.2 is the assignment of the *connection status* feature of *microscope* with the 'connected to microscope' value. The *connection status* feature can only take two qualitative values; 'disconnected' and 'connected to socket'. In order to quantify them, we simply convert 'disconnected' to 0 and 'connected to socket' to 1. The individual score of microscope's *connection status* is 0, when it is unplugged, and 1, when it is plugged into a socket.

Similar is the quantification of the values of *AC switch*'s *state* feature. Since its acceptable values are 'OFF' and 'ON', those are converted into 0 and 1 respectively. In the same fashion, the *attached specimen* feature of *stage* gets the quantified value of 0 when there is no specimen attached and 1 otherwise.

Likewise, we perform the quantification of the values of rest of the features mentioned in Table 4.2.

4.3.7 Normalization

All three examples in the previous subsection concern quantified values which are either 0 or 1. Since individual scores must range from 0 to 1, those values are already normalized, 0 being the *initial* quantified value and 1 the *optimal* one of each of those features (*connection status* of *microscope, state* of *AC switch, attached specimen* of *stage*).

However, we have cases of features that take numerical values not ranging from 0 to 1. Such is the *position* feature of *aperture knob* whose usefulness as a mechanical part of the microscope is to adjust the lamp's light passing through the stage and towards the attached specimen. That feature gets integer values from 0 to 40, corresponding to 41 different positions of the knob from right to left (0 being the position where iris is fully closed and no light passes through, and 40 when iris is fully open). We need to normalize those values into in the [0, 1] interval. This can be done through a *normalization function* with *position* value as argument, which would return 0 in case *position* value is 0 and 1 in case *position* value is 40. A function that serves this purpose is the following:

$$f(x) = \frac{1 + a}{1 + c \cdot (x - 40)^2} - a$$

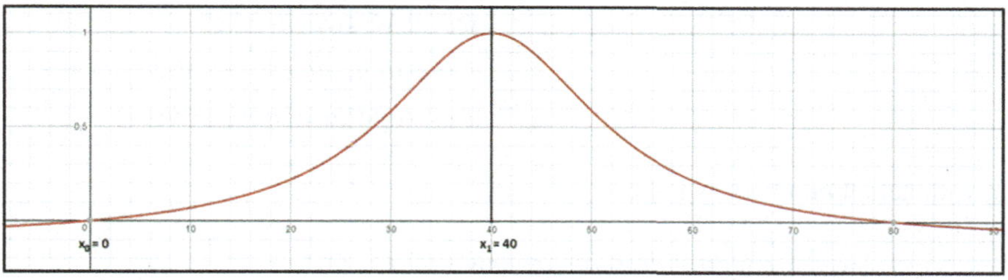

Fig. 4.6 The graph of $f(x) = \frac{1+a}{1+c\cdot(x-40)^2} - a$ for the normalization of aperture knob's position (a = 0.104, c = 0.006)

where $a = 0.104$ and $c = 0.006$.

Figure 4.6 demonstrates the graph of function f.

The normalization of the numerical values of the rest of features is achieved in a similar fashion.

4.3.8 Composite Evaluation

A typical *composite evaluation* that we are dealing with is the case of *microscoping view*'s *clearness* feature, encountered at the last four steps of Table 4.2. As their descriptions suggest, each one of those steps refers to focusing with each one of the objective lenses: 4X, 10X, 40X and 100X.

The *clearness* feature of *microscoping view* is a quantitative measure of the focus achieved with each of the lenses. It is mainly based on *blurriness*, which, taking into account the positions of the various knobs (*coarse focus knob*, *fine focus knob*, *condenser knob*, *left ocular* and *right ocular*), calculates how blurry the final image is in scale from 0 to 1; nevertheless, it depends on the scores for several other "external" features' values, too: those of *AC switch*'s *state*, *light intensity knob*'s *position* and *aperture knob*'s *position*, whose calculation has already been defined.

In order to calculate the value of *clearness* under, say, objective lens 4X, we first need to define the *effect* of *blurriness*. As *blurriness* itself ranges from 0 to 1 with 0 being the optimal value and 1 the worst possible one, we somehow need to "rectify" it so that 0 and 1 would correspond to each other's values, that is, 1 being the optimal and 0 the worst. This is easily achieved by subtracting *blurriness* value from 1:

$$effect(microscoping\ view_{4X}\,|blurriness) = 1 - microscoping\ view_{4X}\,|blurriness$$

Then, the calculation of the composite score for *clearness* is made by multiplying the effect of blurriness with the scores for the afore-mentioned "external" features' values. For example, in case of objective lens 4X, we have:

$$score(microscoping\ view_{4X}\,|clearness) = effect(microscoping\ view_{4X}\,|bluriness)$$

$\cdot\ score(AC\ switch|state)$
$\cdot\ score(light\ intensity\ knob|position)$
$\cdot\ score(aperture\ knob|position)$

or, with actual numbers as indices:

$$x_{13} = effect(microscoping\ view_{4X}\,|bluriness) \cdot x_2 \cdot x_3 \cdot x_4$$

The usefulness of the composite evaluation of *clearness* score is apparent by allowing the "external" values to contribute to it; the higher the score of each one of them, the clearer the final image is and specifically, if they are all equal to 1, score for *clearness* depends exclusively on *blurriness*'s value, while if any of them is 0, score for *clearness* becomes 0 as well.

4.3.9 Success Rate

After the calculation of individual scores x_k comes the *success rate* which combines the former.

As mentioned in the beginning of the section, two different techniques have been used; a Weighted Average and an ANN.

4.3.10 Weighted Average

As its name suggests, the weighted average is the sum of each individual score multiplied by a weight defining its relative importance divided by the sum of those weights:

$$success\ rate \leftarrow \frac{\sum_{k=1}^{n} w_k \cdot x_k}{\sum_{k=1}^{n} w_k}$$

Since individual scores range from 0 to 1, their weighted average does, too. Hence, *success rate* is constrained in the [0, 1] interval.

4.3.11 Artificial Neural Network

Alternatively to defining the success rate as a weighted average, it can also be produced by an ANN.

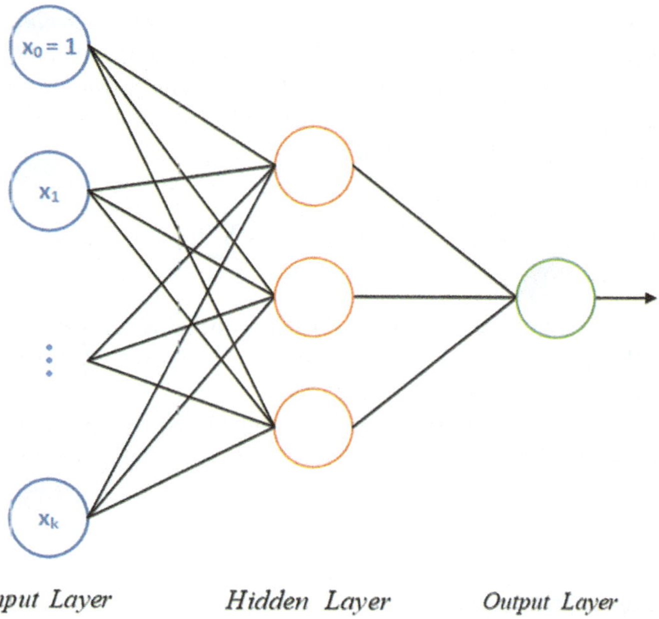

Fig. 4.7 The ANN used for the assessment of user's performance

The ANN we have designed for that purpose is a three-layer one; it contains $k + 1$ input units in the first layer (k units for each individual score x_k and a bias), 3 units in the second (hidden) layer and one output unit in the third layer, while its squashing function for hidden layer and output units is the sigmoid one. The success rate is set to be equal to the normalized value of the output layer unit.[3] Figure 4.7 displays the structure of our ANN.

4.3.12 Penalty Points

In practice, the *success rate* measures the "distance" of the various features' values from their optimal values, that is, the values taken when the procedure in question has been successfully completed. What it does not measure though is the *order* in which those optimal values were achieved; in other words, in which order the necessary steps were made. For example, in our microscoping procedure case, if the user has first turned on the AC switch (Step 2 in Table 4.1) and then connected the microscope plug into the socket (Step 1 in Table 4.1), the *success rate* would be the same as if they made those steps in the correct order (that is, Step 1 before Step 2); however, the overall score should definitely be less.

[3] Unlike the weighted average, the ANN doesn't produce a value within the [0, 1] interval and it therefore needs to be normalized separately.

For this purpose, in case a step is made in the wrong order, a *penalty* is assigned and added to the sum of previously assigned penalties. Specifically, when an i-indexed step is made, the system checks whether the feature values in question at the preceding k-indexed steps ($k < i$) are optimal, and if any of them is not, the overall *penalty points* are updated as follows:

$$penalty\,points \leftarrow old\ penalty + \sum_{k=1}^{i}[u_k \cdot (1 - x_k)] \tag{4.1}$$

where *old penalty* is the sum of all the penalties been assigned so far and u_k the weight for the individual score at Step k (which, in general, is different from the respective w_k weight used for the *success rate*).

As one sees from (4.1), if for an i-indexed step all required previous (k-indexed) steps have preceded it, there is no new assignment of penalty and *penalty points*' value remains the same.

4.3.13 Aggregate Score

Upon the completion of a play session in *evaluation mode*, the user is prompted with *aggregate score* which, like *success rate*, ranges from 0 to 1, but is also affected by other factors which are multiplied with the latter.

First of all, *aggregate score* takes *penalty points* into account, described in the previous sub-section. The more the *penalty points* are, the more we want the *success rate* to decrease.

Second comes *Δtime*, storing the number of seconds passed from the beginning of the session until its ending minus the minimum time required for the completion of the experimental procedure. Like *penalty points*, the greater *Δtime* value is, the less the *success rate* decreases.

Lastly, considering the resetting of instruments as a separate sub-process, we define *resetting rate* as the extent to which the various instrument components have been reset to their original state. In a fashion similar to the definition of *success rate*, we calculate the (unweighted) average of the various individual resetting values. An individual resetting value is the "reversed" score of the respective feature x_i. For example, while the individual score for the *AC switch*'s *state* feature is 0 when *state* is 'OFF' and 1 when *state* is 'ON', its individual resetting value is the opposite, that is, 1 when *state* is 'OFF' and 0 when *state* is 'ON'. Finally, we scale the average into the [0.8, 1] range, as we do not want it to radically decrease the *success rate* value with which it is multiplied, and we get the final *resetting rate* value.

A basic formula of *aggregate score* guaranteeing the afore-mentioned conditions and ranging from 0 to 1 is:

$$aggregate\ score \leftarrow e^{-\frac{penalty\,points}{\beta}} \cdot e^{-\frac{|\delta time|}{\gamma}} \cdot success\ rate \cdot resseting\ rate$$

or:

$$aggregate\ score \leftarrow e^{-\frac{penalty\,points}{\beta}-\frac{|\delta time|}{\gamma}} \cdot success\ rate \cdot resseting\ rate$$

where β and γ are positive constants.

4.4 Machine Learning Algorithms

Having demonstrated the conceptual modeling of Onlabs and its embedded assessment mechanism, it is time to proceed with the presentation of our machine learning techniques used for improving the latter.

4.4.1 Genetic Algorithm for the Weighted Average

We first choose the success rate to be the weighted average.

As already mentioned, in *training mode* we can calibrate the weights of the weighted average with the use of a human expert's feedback and an interactive GA. GAs generally fall into the category of unsupervised learning, using a fixed fitness function for the selection of the members of the new generation. However, our GA is interactive, or, in other words, supervised, meaning that its fitness function is affected by the expert's own evaluation of the user's performance.

We remind that *success rate* is given by the formula:

$$success\ rate \leftarrow \frac{\sum_{k=1}^{n} w_k \cdot x_k}{\sum_{k=1}^{n} w_k} \tag{4.2}$$

where x_k are the individual scores, $k = 1 \ldots n$.

The weights $w_1, w_2, \ldots, w_n$ in *success rate* represent the importance of their respective individual score. We define *weight vector* w as:

$$\vec{w} = (w_1, w_2, \ldots, w_n)^T \tag{4.3}$$

At first, we randomly produce a population of p weight vectors, $\overrightarrow{w^1}, \overrightarrow{w^2}, \ldots, \overrightarrow{w^p}$, or, in short, $\overrightarrow{w^i}$, where $i = 1 \ldots p$. Specializing (4.3) for each weight vector, we get:

$$\overrightarrow{w^i} = \left(w_1^i, w_2^i, \ldots, w_n^i\right)^T \tag{4.4}$$

Now reformulating (4.2) in vector terms, we get:

$$\textit{success rate} \leftarrow \frac{\overrightarrow{w^i} \cdot \vec{s}}{\overrightarrow{w^i}_1}$$

where $\overrightarrow{w^i}_1$ is the first-degree norm of $\overrightarrow{w^i}$ vector, equal to $\sum_{k=1}^{n} w_k^i$.

The first generation of weight vectors is produced by creating a set of weight vectors $\overrightarrow{w^1}, \overrightarrow{w^2}, \ldots, \overrightarrow{w^p}$ with random values from 0 to 100 as their components.

The *fitness function* of our GA is supposed to (probabilistically) choose which weight vectors will survive in the next generation, as a whole or through crossover with other weight vectors. It must therefore take into account how accurately the *success rate* produced by a particular weight vector can approximate the score given by the expert.

After a play session is completed in *training mode*, the computer calculates the various individual scores x_i and the resulting *success rate* while the expert provides his or her own evaluation.

For the individual scores, we create a *score vector* $\vec{x}$, similar to the *weight vector* of (4.3):

$$\vec{x} = (x_1, x_2, \ldots, x_n)^T$$

Assuming for simplicity reasons that one session has only been played, so there is only one *score vector*, $\vec{x}$, and denoting expert's score for this session as $ES(\vec{x})$, we will proceed by defining a generic form of *fitness function*:

$$Fitness_{generic}\left(\overrightarrow{w^i}\right) = 1 - \left| ES(\vec{x}) - \frac{\overrightarrow{w^i} \cdot \vec{x}}{\overrightarrow{w^i}_1} \right| \tag{4.5}$$

The *generic fitness function* gets its maximum value, 1, when the produced *success rate*, $\frac{\overrightarrow{w^i} \cdot \vec{s}}{\overrightarrow{w^i}_1}$, equals the expert's score, and its minimum value when those two quantities are as "far" as possible from each other.

Of course, a single *score vector*, that is, a single play session or in machine learning term, a single training data set, does not suffice for our GA. We therefore play several sessions and are provided with a series of score vectors, $\overrightarrow{x^j}, j = 1 \ldots l$, and their respective *expert scores*, $ES\left(\overrightarrow{x^j}\right)$. Adjusting the generic *fitness function* in (4.5) for each *score vector* $\overrightarrow{x^j}$, we have:

$$Fitness_j\left(\overrightarrow{w^i}\right) = 1 - \left| ES\left(\overrightarrow{x^j}\right) - \frac{\overrightarrow{w^i} \cdot \vec{x}}{\overrightarrow{w^i}_1} \right| \tag{4.6}$$

An obvious *overall fitness function* of weight vector $\overrightarrow{w^i}$ is the average of the various *fitness functions* described in (4.6). Thus, we define:

$$Fitness\left(\overrightarrow{w^i}\right) = \frac{\sum_{j=1}^{l} Fitness_j\left(\overrightarrow{w^i}\right)}{l} = \frac{\sum_{j=1}^{l}\left[1 - \left|ES\left(\overrightarrow{x^j}\right) - \frac{\overrightarrow{w^i}\cdot\overrightarrow{x^j}}{\overrightarrow{w^i}_1}\right|\right]}{l}$$

$$Fitness\left(\overrightarrow{w^i}\right) = \frac{l - \sum_{j=1}^{l}\left|ES\left(\overrightarrow{x^j}\right) - \frac{\overrightarrow{w^i}\cdot\overrightarrow{x^j}}{\overrightarrow{w^i}_1}\right|}{l}$$

$$Fitness\left(\overrightarrow{w^i}\right) = 1 - \frac{\sum_{j=1}^{l}\left|ES\left(\overrightarrow{x^j}\right) - \frac{\overrightarrow{w^i}\cdot\overrightarrow{x^j}}{\overrightarrow{w^i}_1}\right|}{l}$$

where l is the number of the different score vectors, or training examples.

The generic fitness function (hence the overall fitness function, too) that we defined above is negative linear. We also define three alternative generic fitness functions; a negative quadratic, $1 - \left(ES(\vec{x}) - \frac{\overrightarrow{w^i}\cdot\vec{x}}{\overrightarrow{w^i}_1}\right)^2$; a negative exponential, $e^{-\lambda\cdot\left|ES(\vec{x}) - \frac{\overrightarrow{w^i}\cdot\vec{x}}{\overrightarrow{w^i}_1}\right|} e^{-\lambda\cdot\left|ES(\vec{x}) - \frac{\overrightarrow{w^i}\cdot\vec{x}}{\overrightarrow{w^i}_1}\right|}$; and an inverse one, $\frac{1}{\lambda\cdot\left|ES(\vec{x}) - \frac{\overrightarrow{w^i}\cdot\vec{x}}{\overrightarrow{w^i}_1}\right|}$ (λ is a constant greater than 1). Like the negative linear generic fitness function, the three new ones decrease when to the distance between the expert's score and the score produced by the $\overrightarrow{w^i}$ weight vector increases. Among them, the inverse generic function is the one decreasing faster; second comes the negative exponential; third the negative linear; and last the negative quadratic one.

The overall fitness function for the three new generic fitness functions is produced similarly to the first one's. The user may use any of those generic fitness functions.

The new generation to be bred must consist of the same number of weight vectors, that is, p. Whereas a part of them will be "duplicates" of weight vectors from the current generation (*selection* operation), the other part will be "reproduced" by random pairs among those selected weight vectors (*crossover* operation).

At first, we calculate the *selection probability* for each weight vector of the current generation:

$$\Pr\left(\overrightarrow{w^i}\right) = \frac{Fitness\left(\overrightarrow{w^i}\right)}{\sum_{k=1}^{p} Fitness\left(\overrightarrow{w^k}\right)}$$

For the creation of the new generation, we now define a ratio r representing the proportion of replaceable weights (same r for all generations), that is, the number of weight vectors which will not survive into the new one. We then apply the *selection* operation; that is, we select $(1 - r) \cdot p$ weight vectors from the current generation according to their selection probability and “copy” them into the new generation.

At the *crossover* operation, we choose $\frac{r \cdot p}{2}$ pairs of weight vectors from current generation (including those who were selected before) with respect to their *selection probability* to crossover and put their offspring (two for each pair) into the new generation.

Last comes the *mutation* operation. Setting mutation rate to be m, we choose with uniform probability m percent of the weight vectors that have been created in this new generation to be mutated.

We have not defined just one type of mutation but three; doubling of a gene (weight value); halving of a gene; and permutation of two genes within a chromosome (weight vector). The user may choose any one of them, like they choose a generic fitness function.

The same procedure is going on for all generations. The GA halts when it meets our termination condition. We have set the latter to be the production of a particular number of generations.

Our GA abides by Michell’s prototypical GA [27].

4.4.2 *Training the Artificial Neural Network with Back-Propagation*

We now define the *success rate* to be the output given by our ANN shown in Fig. 4.7.

The number of weights of the ANN to be learned are 42 in total (17 weights going from the input units to each 3 of the hidden layer units plus 3 weights going from the latter to the unit in the output layer). The training data sets used in our ANN are no other than the score vectors $\overrightarrow{x^j}, j = 1 \ldots l$, produced by different play sessions, that are also made use of in our GA.

The version of Back-Propagation Algorithm with which we train the ANN is the stochastic gradient descent one for feedforward networks containing two layers of sigmoid units, as described by Michell [27]. As one execution of Back-Propagation for all training data sets does not suffice, we let the algorithm run for several *epochs*.

4.5 Implementation

The implementation of Onlabs has been done in C# programming language under Unity game engine. In the following sub-sections we illustrate each particular mode of Onlabs Machine Learning version 1.0.

Fig. 4.8 Instruction mode: the computer provides the user with instructions of which particular step to follow

4.5.1 Instruction Mode

As mentioned, the instruction mode consists of the computer suggesting by text and voice a particular action and the user being allowed to perform only that. The text appears on a banner at the top as shown in Fig. 4.8.

If the guiding message is not enough, the user may press the hint button on the toolbar (the one with the light bulb icon) to get a new message with more detailed instructions.

4.5.2 Evaluation Mode

In the evaluation mode, the user is allowed to make any action they want while being assessed for their performance in real-time. The real-time evaluation is done in terms both of *success rate* and *penalty points* and shown in a box at the upper right corner (Fig. 4.9).

When the user is finished, they are supposed to press the "final evaluation" button on the toolbar (depicted by a chart icon) to get an overall assessment of their performance in terms of *aggregate score* and all of its components, like the one shown in Fig. 4.10.

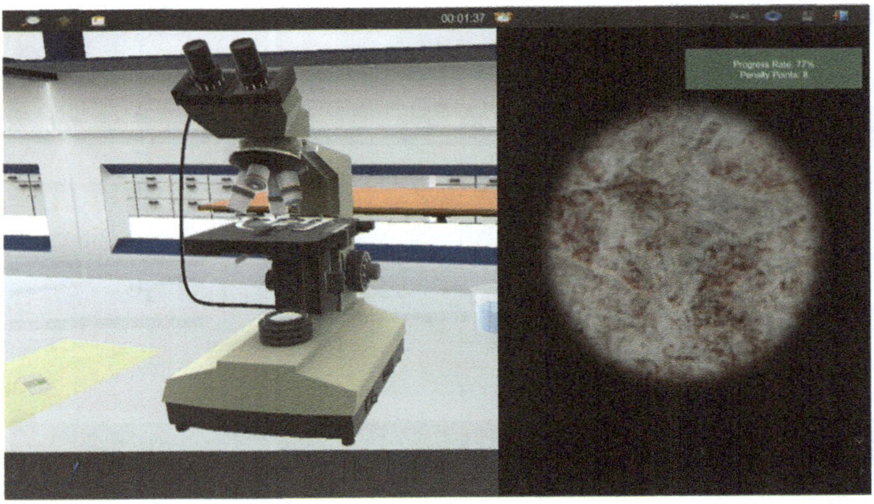

Fig. 4.9 Evaluation mode: the user is being evaluated while playing

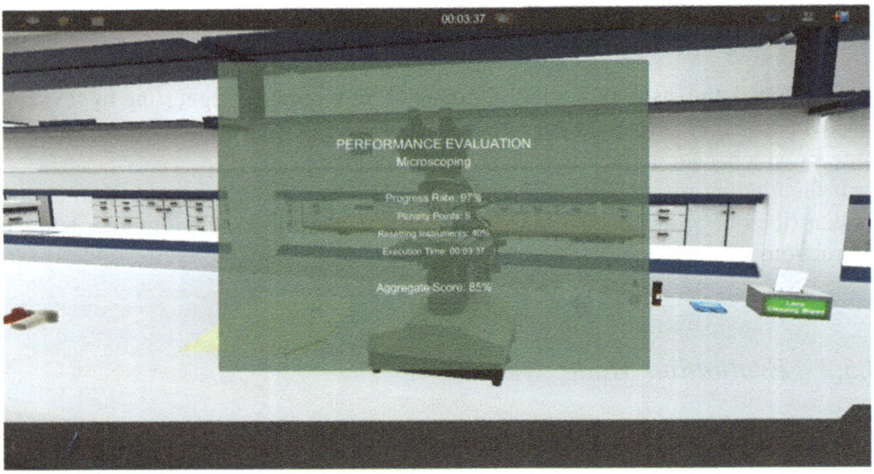

Fig. 4.10 Evaluation mode: upon completing a playing session, the user is prompted with an overall evaluation

4.5.3 Computer Training Mode

Computer training mode consists of two separate sub-modes; training data collection, where the user plays a session and a human expert provides their own evaluation of the former's *success rate* (not *penalty points*), and machine learning, where the rater, that is, any of the computer's embedded assessment mechanisms, is trained with the use of the training data already collected.

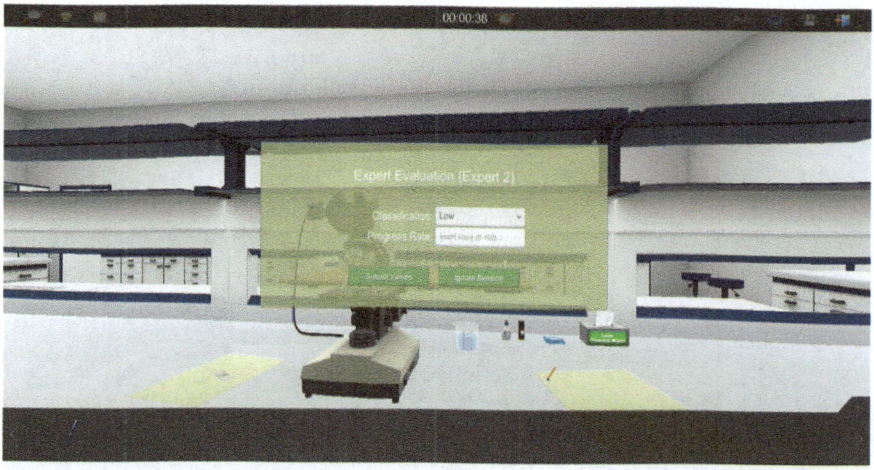

Fig. 4.11 Computer training mode/training data collection: when collection of training data is performed, the expert is asked to provide his or her own evaluation upon the completion of each play session

4.5.4 *Training Data Collection Sub-mode*

When the user completes a play session of *training mode*, an evaluation panel shows up in which the expert is asked to submit their own assessment from 0 to 100, later scaled to 0-1 range and saved along with the individual scores achieved in this session. The individual scores and the expert's evaluation consist of a single training data set. New play sessions produce new training data sets, which all together are to be used by our machine learning algorithms. The expert is also expected to choose a 'Low', 'Medium' or 'High' performance classification for the session in accordance to their evaluation; for example, a 'Low' classification is consistent with an expert's evaluation of 10%, 'Medium' with 40% and 'High' with 75%. The expert's evaluation panel is shown in Fig. 4.11.

4.5.5 *Machine Learning Sub-mode*

When enough training data have been collected, the user may well start the rater training procedure with the GA on the Weighted Average or the Back-Propagation on the ANN.

When executing the GA on the Weighted Average, the user is asked to define the number of population members (p), the crossover rate (r), the mutation rate (m), the number of generations for the GA to run as well as the generic fitness function and the mutation method to be used (Fig. 4.12). When executing the Back-Propagation on the ANN, they are asked for the bias and the epochs instead (Fig. 4.13).

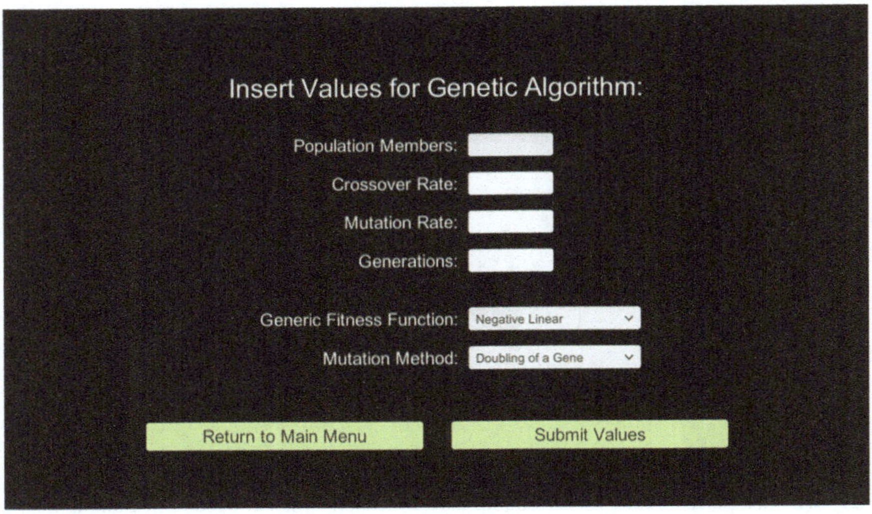

Fig. 4.12 The user is asked to enter the values for GA training parameters

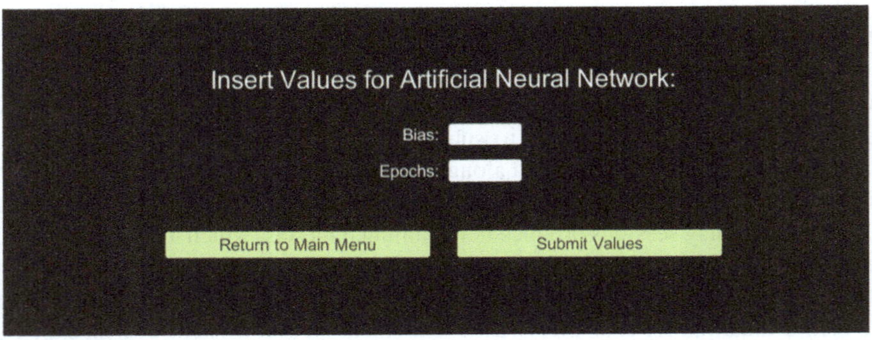

Fig. 4.13 The user is asked to enter the values for ANN training parameters

When the training is completed, the user is prompted with a table containing the mean squared errors for various data groupings. Figures 4.14 and 4.15 show the respective tables for ANN and GA trainings with various parameter values.

As illustrated in Fig. 4.14, in the case of GA, the similarity between the original intuitional weight vector and the produced weight vector is also shown. Moreover, in both machine learning cases, the user is asked whether they want to save the produced weights in order to be used in the respective scoring mechanism of the Evaluation Mode. Moreover, apart for the rather brief table with the mean squared errors shown on screen, a more detailed Comma-Separated Value (.csv) file is produced containing the evolution of those mean squared errors through every generation or epoch.

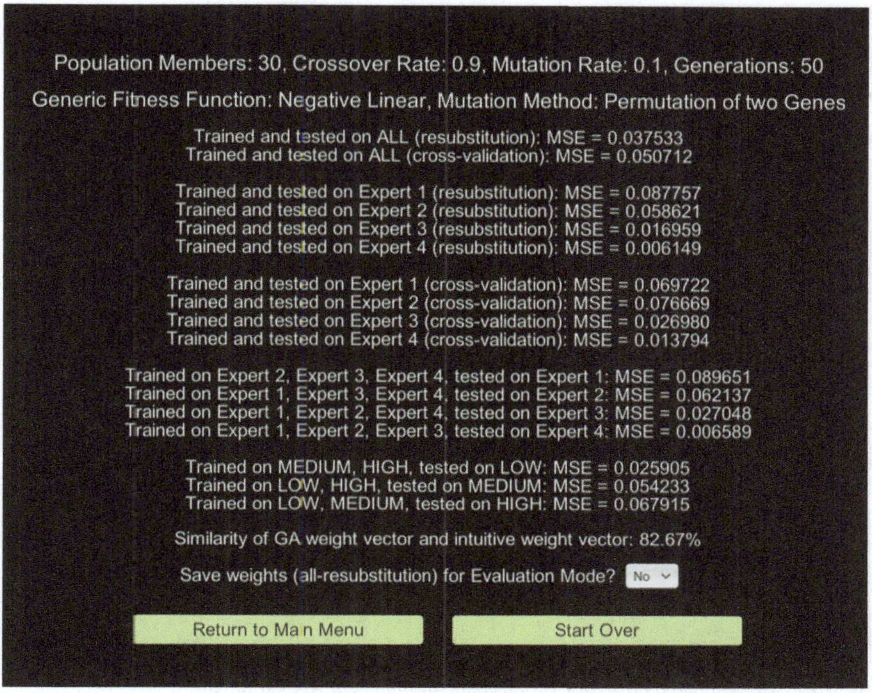

Fig. 4.14 Mean squared errors for various data groupings in GA training

4.6 Training-Testing Process and Results

In the previous section, we thoroughly described the machine learning algorithms used in the rater training sub-mode. In this section, we will go through the training and testing process and the results in terms of the Mean Squared Error (MSE) produced by various training sessions. But first, we need to make reference of our collected training data.

4.6.1 *Training Data*

As we mentioned earlier in the chapter, every training data set consists of the individual scores achieved by a human user in a play session plus the estimation of their achieved success rate given by a human expert, along with the name of the expert and the 'Low', 'Medium' or 'High' classification of the session in terms of the user's performance.

For the purpose of our research, we asked 4 biology experts at Hellenic Open University to evaluate 60 play sessions by us. In particular, each expert evaluated 15

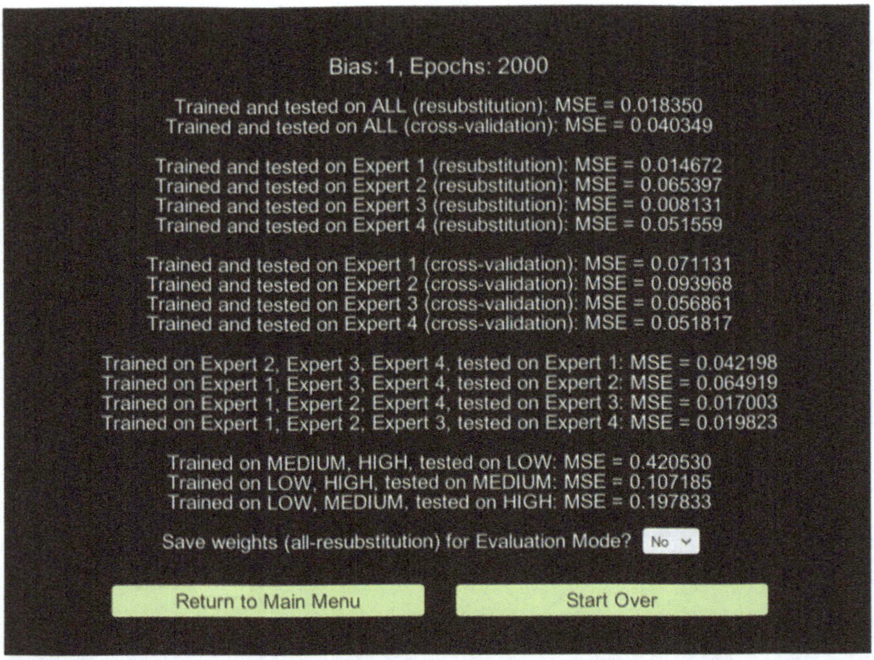

Fig. 4.15 Mean squared errors for various data groupings in ANN training

play sessions. Among those 15 sessions, 5 were classified as 'Low', 5 as 'Medium' and the other 5 as 'High'. Therefore, in total we have 20 sessions classified as 'Low', 20 as 'Medium' and 20 as 'High'.

4.6.2 Training and Testing on Various Data Set Groups

In order to do the training and testing, we form various groups of our data sets.

In the first, "optimistic" case, we apply resubstitution on all data sets, that is, train our system and test training results on them getting the respective MSE. Then we apply cross-validation on all data sets, that is, train on some of the data sets and test the results on the rest of them.

The groups of data sets used for training and testing in cross-validation are formed like in the following fashion: From each expert, we choose a data set classified as 'Low', another one classified as 'Medium' and a final one classified as 'High'; in total we have a group of 12 data sets, equally divided among the 4 experts and the 3 classifications. Then we likewise choose another 3 data sets from each expert (not the ones chosen before) and we form a second group of 12 data sets with the aforementioned properties. We repeat this kind of selection 3 more times and we finally have 5 groups of 12 data sets while each group contains one 'Low', one 'Medium' and one 'High' data set from each expert. Thus, in cross-validation for all data sets, we train on 4 groups chosen among them and test on the respective 5th one in all possible combinations and calculate their average of MSEs.

Afterwards, we perform resubstitution and cross-validation within every expert. In a similar way to the previous grouping, we form 5 groups from each expert, each one of which containing a 'Low', a 'Medium' and a 'High' data set; again, we each time perform training on 4 groups of them and testing on the respective 5th one and finding their average of MSEs.

Then comes cross-validation among all experts, meaning training on all data sets corresponding to 3 experts and testing on the respective 4th, in all combinations.

Finally, we have cross-validation among classifications, that is, training on the data sets falling under any of 2 chosen classifications and testing on the respective 3rd, again in all possible combinations.

4.6.3 Genetic Algorithm Results

We tried numerous possible combinations of generic fitness functions and permutation methods with various different values for the number of population members, the crossover rate, the mutation rate and the number of generations.

Convergence was achieved for one particular case, that of the inverse generic fitness function with halving of a gene as mutation method. There we have very small convergence of around 0.004 and 0.003 magnitude for resubstitution and cross-validation respectively, as shown in Fig. 4.16.

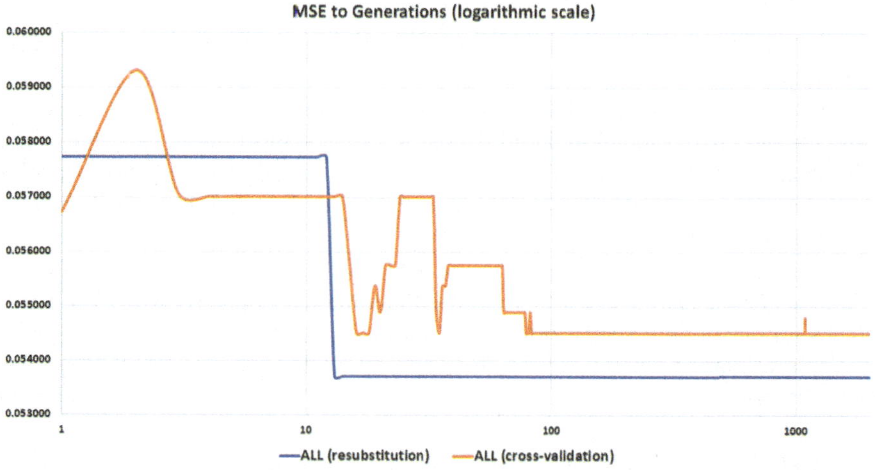

Fig. 4.16 MSE to Generations graphs (logarithmic scale) for GA—resubstitution and cross-validation on all data sets (population members: 100, crossover rate: 0.9, mutation rate: 0.01, generations: 2000, generic fitness function: inverse, mutation method: halving of a gene)

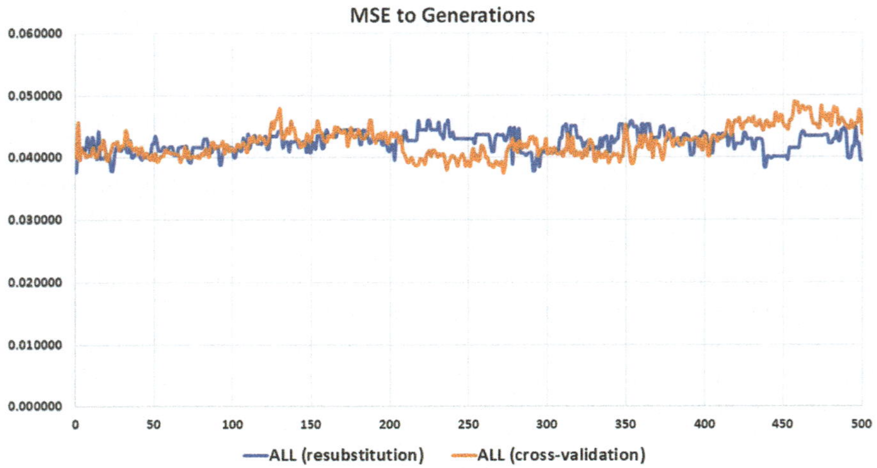

Fig. 4.17 MSE to Generations graphs for GA—resubstitution and cross-validation on all data sets (population members: 100, crossover rate: 0.9, mutation rate: 0.01, generations: 2000, generic fitness function: negative linear, mutation method: permutation of two genes)

In the rest of the cases, no convergence was achieved. A typical behavior of no convergence is the one of negative linear generic fitness function with permutation of two genes as the mutation method, shown in Fig. 4.17.

4.6.4 Artificial Neural Network Training Results

In contrast with the GA, in the ANN training case the convergence is achieved in most cases of groups of data sets.

Figure 4.18 shows the MSE to Epochs graphs (bias: 1, epochs: 2000) for resubstitution and cross-validation on all data sets; in the resubstitution case, the graph converges indefinitely while in the cross-validation case, the convergence becomes maximum in approximately the 400th epoch and starting slightly to diverge from that point forward.

The same behavior is exhibited in the resubstitution and cross-validation cases within a single expert's data sets. Figure 4.19 shows the respective MSE to Epochs graphs for one of our experts for the same bias and epochs.

Cross-validation among experts also converge within the first 400-800 epochs, then slightly diverge and after some point start converging again, as shown in Fig. 4.20.

On the contrary, no reliable convergence or no convergence at all is achieved in the cross-validation among different classifications, as shown in Fig. 4.21.

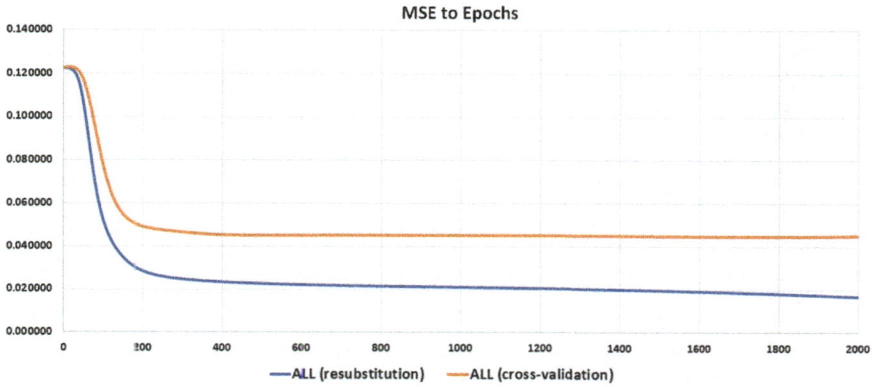

Fig. 4.18 MSE to Epochs graph for ANN training—resubstitution and cross-validation on all data sets (bias: 1, epochs: 2000)

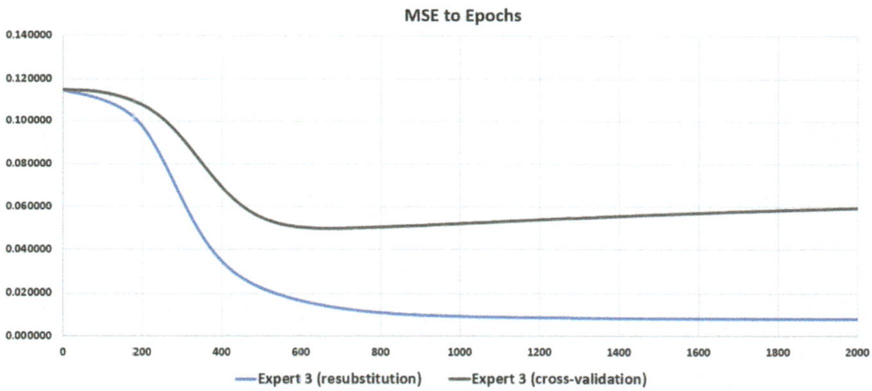

Fig. 4.19 MSE to Epochs graph for ANN training—resubstitution and cross-validation within one of our expert's data sets (bias: 1, epochs: 2000)

4.7 Conclusions

Onlabs is an adventure game-like 3D virtual lab developed at Hellenic Open University for its biology students to be trained before they use the on-site lab. Its recent stable version 2.1.2 includes two experimental procedures, those of microscoping of a test specimen and 10X TBE solution preparation. Apart from its simple modes of free experimentation and instruction, Onlabs also offers the mode of evaluation, where the user is assessed for his or her performance and the mode of training contained in its separate Machine Learning version 1.0, where the assessment mechanisms are reconfigured with the use of an interactive GA on the weighted average and Back-Propagation on the ANN that we have designed and implemented for that purpose. For the moment, training mode concerns only the microscoping procedure.

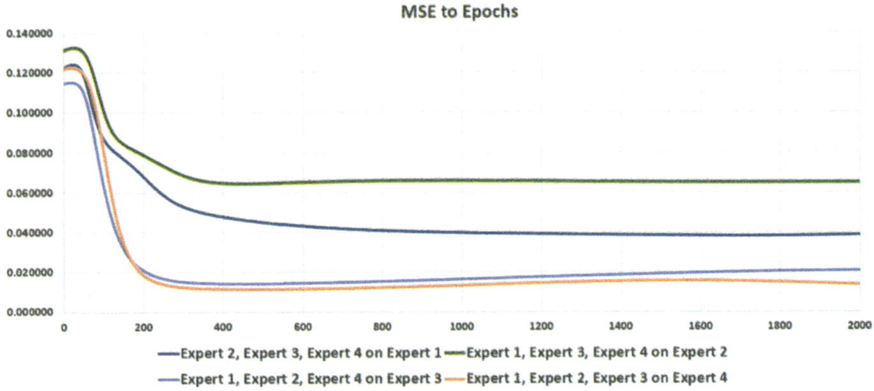

Fig. 4.20 MSE to Epochs graph for ANN training—cross-validation among experts (bias: 1, epochs: 2000)

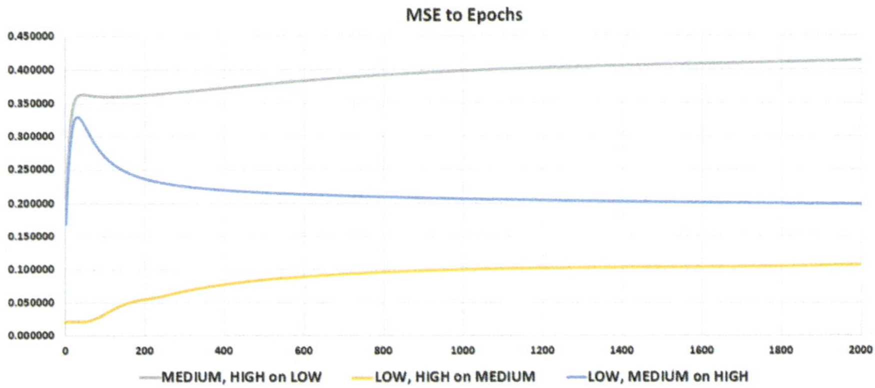

Fig. 4.21 MSE to Epochs Graph for ANN training—cross-validation among classifications (bias: 1, epochs: 2000)

One of our short-term future goals is the expansion of training mode so that it will include the 10X TBE solution preparation procedure as well as the electrophoresis one implemented in Onlabs beta version 3.0. Recently, we also completed the design and implementation of reinforcement learning algorithms for the NPC's training for microscoping procedure and we are going to expand it for the other procedures, too.

Development-wise, we are planning to equip Onlabs with more experimental procedure as well as to deploy new virtual labs for Chemistry and Physics.

Acknowledgements This research has been co-financed by the Operational Program "Human Resources Development, Education and Lifelong Learning" and is co-financed by the European Union (European Social Fund) and Greek national funds.

References

1. J. Brooke, SUS—A quick and dirty usability scale, in *Usability Evaluation in Industry*, ed. by P.W. Jordan, B. Thomas, I.L. McClelland, B.A. Weerdmeester (Taylor & Francis, London, UK, 1996), p. 7
2. V. Zafeiropoulos, D. Kalles, A. Sgourou, Learning by playing: development of an interactive biology lab simulation platform for educational purposes, in *Experimental Multimedia Systems for Interactivity and Strategic Innovation* (2016), pp. 204–221
3. V. Zafeiropoulos, D. Kalles, A. Sgourou, A. Kameas, Adventure-style serious game for a science lab, in *Open Learning and Teaching in Educational Communities*, ed. by C. Rensing, S. de Freitas, T. Ley, P.J. Muñoz-Merino. (Springer International Publishing, 2014), pp. 538–541
4. E. Paxinou, A. Karatrantou, D. Kalles, C. Panagiotakopoulos, A. Sgourou, A 3D virtual reality laboratory as a supplementary educational preparation tool for a biology course. Eur. J. Open, Dist. E-learn. **21** (2018)
5. E. Paxinou, V. Zafeiropoulos, A. Sypsas, C. Kiourt, D. Kalles, Assessing the impact of virtualizing physical labs. arXiv:1711.11502 [cs] (2017)
6. E. Amir, P. Doyle, Adventure games: a challenge for cognitive robotics. Am. Assoc. Arti. Intell. 8 (2002) (www.aaai.org)
7. B. Hlubocky, E. Amir, Knowledge-gathering agents in adventure games (2004)
8. A.L. Thomaz, C. Breazeal, Reinforcement learning with human teachers: evidence of feedback and guidance with implications for learning performance, in *Proceedings of the 21st National Conference on Artificial Intelligence*, vol. 1 (AAAI Press, 2006), pp. 1000–1005
9. A.L. Thomaz, C. Breazeal, Teachable robots: understanding human teaching behavior to build more effective robot learners. Arti. Intell. **172**, 716–737 (2008). https://doi.org/10/d49w2c
10. P. Ammanabrolu, M. Riedl, Playing text-adventure games with graph-based deep reinforcement learning, in *Proceedings of the 2019 Conference of the North American Chapter of the Association for Computational Linguistics: Human Language Technologies*, vol. 1 (Long and Short Papers) (Association for Computational Linguistics, Minneapolis, Minnesota, 2019), pp. 3557–3565. https://doi.org/10/gf9fzz
11. K. Narasimhan, T. Kulkarni, R. Barzilay, Language understanding for text-based games using deep reinforcement learning. arXiv:1506.08941 [cs]. (2015)
12. B. Kostka, J. Kwiecien, J. Kowalski, P. Rychlikowski, Text-based adventures of the golovin AI agent, in *2017 IEEE Conference on Computational Intelligence and Games (CIG)* (2017), pp. 181–188. https://doi.org/10/gf9fz2
13. D. Robitzski, A neural network dreams up this text adventure game as you play. https://futurism.com/text-adventure-game-neural-network. Last accessed 5 Oct 2019
14. M.S. Robbins, Using neural networks to control agent threat response, in *Game AI Pro 360: Guide to Tactics and Strategy*, ed. by S. Rabin. (CRC Press, 2019), p. 242
15. D. Charles, C. Fyfe, D. Livingstone, S. McGlinchey (eds.), Biologically inspired artificial intelligence for computer games. IGI Global (2008). https://doi.org/10.4018/978-1-59140-646-4
16. J.J. Luo, An exploration of neural networks playing video games. https://towardsdatascience.com/an-exploration-of-neural-networks-playing-video-games-3910dcee8e4a. Last accessed 26 Sept 2019
17. A. Arora Using GAs to automate the chrome dinosaur game (Part 2), https://heartbeat.fritz.ai/using-genetic-algorithms-to-automate-the-chrome-dinosaur-game-part-2-1c0007334297. Last accessed 26 Sept 2019
18. V.G. de Mendonça, C.T. Pozzer, R.T. Raittz, A framework for GAs in games. Presented at the (2008)
19. M. Martin, Using a GA to create adaptive enemy AI, https://www.gamasutra.com/blogs/MichaelMartin/20110830/90109/Using_a_Genetic_Algorithm_to_Create_Adaptive_Enemy_AI.php. Last accessed 26 Sept 2019

20. D. Gong, J. Yuan, X. Ma, Interactive GAs with large population size, in *2008 IEEE Congress on Evolutionary Computation* (IEEE World Congress on Computational Intelligence) (2008), pp. 1678–1685. https://doi.org/10/fg8fhb
21. W.M. Spears, K.A. De Jong, Using GAs for supervised concept learning, in *Proceedings of the 2nd International IEEE Conference on Tools for Artificial Intelligence* (1990), pp. 335–341. https://doi.org/10/fkkp8g
22. X. Sun, D. Gong, W. Zhang, Interactive GAs with large population and semi-supervised learning. Appl. Soft Comput. **12**, 3004–3013 (2012). https://doi.org/10/f34pf6
23. X. Sun, J. Ren, D. Gong, Interval fitness interactive GAs with variational population size based on semi-supervised learning, in *Advances in neural networks—ISNN 2010*, ed. by L. Zhang, B.-L. Lu, J. Kwok (Springer, Berlin Heidelberg, 2010), pp. 288–295
24. S. Russell, P. Norvig, *Artificial Intelligence: A Modern Approach* (Pearson Education, Upper Saddle River, NJ, USA, 2003)
25. D. Poole, A. Mackworth, *Artificial Intelligence: Foundations of Computational Agents* (Cambridge University Press, New York, NY, USA, 2010)
26. E. Yourdon, *Modern Structured Analysis* (Yourdon Press, Upper Saddle River, NJ, USA, 1989)
27. T.M. Mitchell, *Machine Learning* (McGraw-Hill Inc, New York, NY, USA, 1997)

Chapter 5
Exploiting Semi-supervised Learning in the Education Field: A Critical Survey

Georgios Kostopoulos and Sotiris Kotsiantis

Abstract Educational Data Mining and Learning Analytics are two interrelated and fast-growing research fields with a view to extracting meaningful information from educational data and enhancing the quality of learning. Predicting student learning outcomes is one of the most significant problems facing these fields. Addressing effectively a predictive problem comprises the training of a supervised learning algorithm on a given set of labeled data. The difficulty of obtaining a sufficient amount of labeled data in many practical problems has resulted to the development of new machine learning approaches which are generally referred to as Weakly Supervised Learning. Semi-Supervised Learning and Active Learning constitute the main components of Weakly Supervised Learning with a view to exploiting a small pool of labeled examples together with a large pool of unlabeled ones in the best possible manner for building highly accurate and robust learning models. Over the last few years, a plethora of Semi Supervised Learning algorithms have been developed and implemented with great success for solving a variety of problems in many scientific fields, among which the education field as well. Following up on recent research, the main purpose of the present study is to provide a comprehensive review on the applications of Semi Supervised Learning in the fields of Educational Data Mining and Learning Analytics. The analysis of the relevant studies reveals that Semi Supervised Learning constitutes a very effective tool for both early and accurate prognosis of student learning outcomes anticipating better results from traditional supervised methods.

Keywords Educational data mining · Learning analytics · Semi-supervised learning · Student learning outcomes · Prediction

---

G. Kostopoulos · S. Kotsiantis (✉)
Educational Software Development Laboratory (ESDLab), Department of Mathematics, University of Patras, Patras, Greece
e-mail: sotos@math.upatras.gr

G. Kostopoulos
e-mail: kostg@sch.gr

G. A. Tsihrintzis et al. (eds.), *Advances in Machine Learning/Deep Learning-based Technologies*, Learning and Analytics in Intelligent Systems 23,
https://doi.org/10.1007/978-3-030-76794-5_5

5.1 Introduction

One of the greatest advances in the field of education over the last two decades is the integration of Information and Communication Technologies (ICT). Nowadays, ICT is an integral part of the educational process offering teachers and students a variety of interactive learning environments to support teaching and learning and enhance the quality of education [1]. As a result, large volumes of data are constantly produced and stored on databases and information systems regarding students' achievements, learning behavior, and online activity within the institutional Learning Management Systems (LMS). The necessity to analyze different types of educational data and retrieve valuable information from them has contributed on the genesis of two closely interrelated and fast-growing research fields: Educational Data Mining (EDM) and Learning Analytics (LA) [2].

EDM focuses primarily on the development and implementation of data mining methods on educational data arising from a variety of learning environments for solving essential educational problems [3], while LA is mainly centered around the learning process, exploiting data analysis to reinforce the decision-making procedures [4]. However, irrespectively of the method used to handle an educational problem, both scientific fields share common goals: gain understanding about students' learning behavior, enhance learning process and improve the quality of education [5].

One of the most familiar EDM and LA methods is prediction, where the goal is to build a machine learning model training a supervised algorithm (e.g. a decision tree) on a set of labeled data and subsequently applying it to predict unknown or future events [4]. For example, predicting whether a student will drop out or graduate from a university unit or a course. Unfortunately, in many practical problems it is difficult, if not impossible, to obtain a sufficient amount of labeled training data due to the high cost of labeling. This barrier has resulted to the development of new machine learning approaches which are generally referred to as Weakly Supervised Learning [6]. Semi-Supervised Learning (SSL) and Active Learning (AL) constitute the main components of Weakly Supervised Learning with a view to exploiting a small pool of labeled examples together with a large pool of unlabeled ones for building highly accurate and robust learning models.

Over the last few years, a plethora of SSL methodologies and algorithms have been introduced and employed with great success for solving a variety of machine learning problems in many scientific fields. As a result, a great deal of studies has emerged recently concerning the implementation of SSL methods in the educational field demonstrating very promising results compared to the supervised ones [7]. However, there is a luck of a literature review summarizing these studies and introducing the most promising ones.

To this end, the major purpose of the present study is to provide a comprehensive review of SSL applications in the area of EDM and LA. In addition, we intend to propose a taxonomy based on the learning outcome predicted in them. Two main research questions guide our study:

Q1: What learning outcomes have been predicted when implementing SSL methods in educational data?

Q2: Which are the gains of using SSL methods for predicting learning outcomes of students?

The remaining of the study is organized as follows: Sect. 5.2 is devoted to the SSL approach and presents two representative self-labeled methods: self-training and co-training. Section 5.3 discusses several studies investigating the effectiveness of SSL methods in the education field mainly for prediction problems. Finally, Sect. 5.5 concludes the study considering new paths for future work.

5.2 Semi-supervised Learning

SSL is a typical example of learning from both labeled and unlabeled data. Since in many real world applications unlabeled data is abundant and easy to obtain while the labeling cost is sufficiently high, SSL has received increasing attention among scientists in recent years. The impressive development of SSL has been also motivated by its practical value in building highly accurate learning models and its theoretical value in understanding machine and human learning [8]. The main concept behind SSL is to automatically utilize a small pool of labeled examples together with a large pool of unlabeled ones without human intervention and build powerful learning models [6]. Various methodologies embracing the core idea of SSL have been originally implemented and successfully applied for solving a wide range of tasks, such as web mining, text mining, image processing, information retrieval and bioinformatics [9].

Depending on the nature of the output attribute, SSL appears in two main forms: Semi-Supervised Classification (SSC) for a discrete output attribute and Semi-Supervised Regression (SSR) for a real-valued one [10]. Formally, the SSL setting (Fig. 5.1) can be defined as follows [6]:

Given a training dataset $D = L \cup U$, where $L = \{(x_1, y_1), (x_2, y_2), \ldots, (x_k, y_k)\}$ consists of k labeled examples and $U = \{x_1, x_2, \ldots, x_m\}$ consists of m unlabeled examples, with $m \gg k$, $x_i \in X$ and $y_i \in Y$, we want to learn a function $f : X \rightarrow Y$ for predicting either the class label (SSC) or the value (SSR) of y for any new vector example $x \in X$ (inductive SSL) or for any vector example $x \in U$ (transductive SSL) [11].

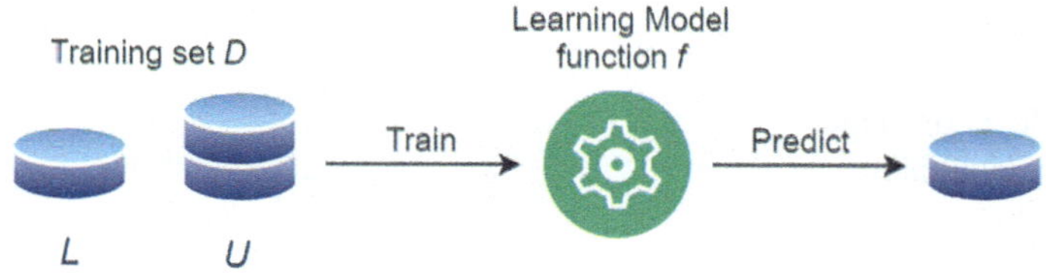

Fig. 5.1 The SSL setting at a glance

Self-training [12] and co-training [13] are typical SSC algorithms which harvest the full benefits from the information hidden in unlabeled data for improving the learning performance. What is more, these algorithms laid the ground for further developing more sophisticated and efficient SSC and SSR techniques.

Self-training or single-view weakly supervised learning [14] is a simple and commonly used semi-supervised algorithm which aims to obtain a progressively enlarged labeled dataset for learning the function f through an iterative learning process. More precisely, a supervised algorithm h (e.g. a decision tree) is trained on the labeled dataset L and subsequently is applied in U. The most confident predicted examples (x, y^*) with $x \in U$ are selected and integrated into L, h is retrained, and the procedure is repeated until some stopping criterion is met. Self-training is a wrapper self-labeled method whose effectiveness strongly depends on the supervised algorithm h and the labeling confidence at each iteration of the learning process [15] which, in fact, determines the quality of model h [11].

Co-training is a representative multi-view semi-supervised algorithm, which assumes that each example can be represented by two distinctive and independent set of features, which are known as views. Two supervised algorithms h_1 and h_2 are separately trained on each view V_1 and V_2 of the labeled dataset L, and the most confident predicted examples of h_1, h_2 on U are integrated into the training set of the other. Finally, h_1, h_2 are retrained on the augmented labeled dataset and the procedure is repeated until some stopping criterion is met or U is empty.

Since the assumption about the existence of two feature views can hardly be met in most real-world applications, a number of co-training modifications have been designed and implemented, such as tri-training [16], Rasco [17] and Rel-Rasco [18], to name a few.

5.3 Literature Review

Over the last few years, a growing body of research has investigated the effectiveness of SSL methods in the education field primarily for prediction problems. As a result, a great deal of studies has emerged demonstrating very promising results compared to the traditional supervised methods [7]. These studies can be classified according to the following three criteria:

Criterion 1: The predicted learning outcome:

- Performance.
- Dropout.
- Grade level or range.
- Grade point value.
- Others.

Criterion 2: The level of education:

- University.
- Traditional.
- Open.
- Secondary education.
- High school.
- Middle school.

Criterion 3: The machine learning task:

- Classification.
- Regression.

According to the first criterion, the studies are classified into five groups and discussed in the following subsections. The main interest of these studies relates to classification problems (binary or multiclass) such as the prediction of student performance, dropout, and grade level, while they are focusing, in principle, on students enrolled in university courses.

5.3.1 *Performance Prediction*

Predicting student performance is one of the main interests in the area of EDM and LA [19]. We consider two aspects of the student performance task: predicting whether a student will pass or fail a unit or a course and grouping students in more than two classes according to their performance.

The efficiency of several SSL algorithms was examined in [20] for identifying undergraduate students at risk of failure in the final examinations of a one-year unit course. Self-training, co-training with random feature split, tri-training, tri-training with editing, democratic co-learning, Rasco and Rel-Rasco algorithms were applied employing four classifiers as base learners: the Naïve Bayes (NB) discriminative classifier, the C4.5 decision tree, the k-Nearest Neighbor instance-based algorithm and the Sequential Minimization Optimization (SMO) algorithm. SSL algorithms proved to be quite effective for the early prediction of low performers. Amidst these methods, tri-training with three C4.5 decision trees as base classifiers (different parameters were used for ensuring the diversity between classifiers) prevailed not only over the rest SSL algorithms, but also over the C4.5 supervised classifier. A tri-training variant was applied in [21] for the same task. Three Random Forest (RF) classifiers were incorporated as base learners, while their predictions were fed sequentially into a self-training algorithm for making the final estimation.

A variant of self-training embedding the RF classifier as base learner was utilized in [22] for predicting the study status of university students at the end of their course. In addition, the RF classifier comprised three decision trees which were constructed from random subsets of features, thus operating in a tri-training mode. The proposed method proved to be sufficiently effective for detecting students at risk of failing since it outperformed familiar supervised and semi-supervised methods. For the same task, a variant of tri-training was developed using three RF classifiers as base learners [23]. Each forest consisted of 300 decision trees, while two subsets of features were randomly sampled for splitting a node. Besides that, an imputation method was adopted for exploiting incomplete data, in accordance with which the mean of the k=3 nearest neighbors of an example was used for replacing unregistered grades of students before providing the examples to the base learner for training. The experimental results showed improvement in prediction accuracy using different ratios of labeled data (12.5, 20, 25 and 50%). In a similar study, a simplified version of the same methodology was applied incorporating the self-training algorithm [24]. Two variants of the proposed method were proposed, an iterative and a sequential, while moreover the percentage of unlabeled examples in the experiments carried out ranged between 50 and 95%. The experimental results showed a slight decrease of the accuracy as the percentage of unlabeled examples was increased, while the iterative algorithm proved to be the more accurate.

Transfer learning and co-learning were combined in [25] for predicting the study status of students enrolled in four-years degree university programs. Hence, a learning model was build based upon data from a program P_1 and accordingly was applied for predicting student study status in another similar program P_2. Finally, the co-training algorithm was employed for classifying students in three classes: graduating, studying and stop studying.

Quite recently, a co-training method was formulated for the early prognosis of students' performance in distance higher education based upon the existence of two totally different and independent feature views: academic achievements and LMS activity [7]. A plethora of experiments were carried out testing the effectiveness of the SSL method against a variety of self-training and co-training variants, and the co-Forest algorithm. Additionally, three different experimental scenarios were adopted considering three different labeled ratios in the training dataset (2.5%, 10% and 15%). The co-training method performed better than the rest SSL algorithms regardless of the labeled ratio and the base learner employed in terms of accuracy and F-score. It is worth noting the supremacy of the proposed method over familiar supervised classification methods such as RF, NB, k-NN (k=5) and Extra Trees.

5.3.2 Dropout Prediction

Student dropout is a major concern in higher education, and especially in open universities [26]. In this regard, the efficiency of the SSL approach was examined in [27] for predicting students who are likely to drop out of a course in distance higher education. A plethora of experiments were carried out in three steps dividing the academic year into three consecutive time periods. Several SSC algorithms were formed employing two familiar classifiers as base learners (i.e. C4.5 and NB). The experimental results revealed that tri-training and self-training produced very reliable predictions, with an accuracy measure ranging from 71.74 to 76.73% before the middle of the academic year.

5.3.3 Grade Level Prediction

Several SSC algorithms were applied in [28] for inferring the class grade (i.e. poor, good, very good, excellent) of high school students in the examinations of the mathematics module at the end of the academic year. The experimental results indicated that self-training, tri-training and co-training (the NB algorithm was used as base classifier) prevailed with an accuracy value ranging from 64.41 to 67.35% at the middle of the academic year. Based upon the findings of this research, three ensemble SSC methods were implemented in [29] deploying self-training, co-training and tri-training respectively as base learners. The predictions of three classifiers (C4.5, SMO and Logistic Model Tree (LMT)) were combined by majority voting in each method, thereby producing the final output of the ensemble. Several experiments were carried out for three different labeled ratios (10%, 20% and 30%) highlighting the predominance of the ensemble-based SSL methods for the accurate prediction of students' class grade.

The effect of social influence at predicting students' grade level based on their campus social behavior was studied in [30]. For this purpose, a social network was created for each student from digital records based on his/her campus daily activities. In addition, it was found that students with strong social ties tend to have similar academic performance. Finally, a Label Propagation algorithm was trained on different labeled ratios (16, 32, 48, 64, 80%) and subsequently applied for predicting the final grade level of students (four classes of the Grade Point Average were used).

5.3.4 Grade Point Value Prediction

Importantly, we distinguish three studies dealing with the development and deployment of SSR methods for prediction tasks. An extension of the supervised multi-task regression in the semi-supervised setting was proposed in the first study for predicting examination grades of secondary students [31]. More precisely, for each learning task a Gaussian Process (GP) was incorporated as the base regressor under the assumption that there is a common prior on the kernel parameters for all tasks. Following this, the unlabeled data were utilized with an appropriate adjustment of the Radial Basis Function (RBF) kernel function. Furthermore, pairwise information was adopted for improving the prediction accuracy by exploiting the information hidden in unlabeled data. The experimental outputs revealed the supremacy of the method in terms of normalized Mean Square Error (MSE).

An ensemble-based SSR algorithm was introduced in [10], termed Multi-Scheme Semi-Supervised Regression Approach (MSSRA), for grade prediction of undergraduate students in the final examinations of a one-year unit course. Three k-NN regressors were deployed in a self-training scheme for augmenting the labeled dataset through an iterative learning process utilizing the unlabeled data. Finally, a RF regressor was employed for building the regression model. The proposed SSR algorithm prevailed over familiar regression methods (i.e. RF, Linear Regression (LR),

k-NN, M5 Rules, M5 Model Tree and SMOreg) in terms of four metrics: Mean Absolute Error (MAE), Relative Absolute Error (RAE), Root Mean Squared Error (RMSE) and Pearson Correlation Coefficient (PCC).

Very recently, a multi-view semi-regression algorithm was implemented for predicting the grades of undergraduate students in the final examinations of a one-year distance learning course [32]. To this end, a variant of the COREG algorithm (CO-training REGressors) [33] was employed, whereas the prediction was made for two time-periods before the middle of the academic year. Additionally, the study explored the influence of the input attributes over the target one, thus producing a plethora of explainable diagrams regarding their impact on the output of the SSR model. The experimental results uncovered that the best performers were students who earned high grades in the two compulsory written assignments during the first semester.

5.3.5 Other Studies

In addition, there are a few studies related to automatic grading, problem-solving performance, concept extraction and outcomes mining. A semi-supervised clustering method was developed in [34] for automatically grading short answers of students attending a Massive Open Online Course (MOOC). According to this method, similar answers should be grouped together since they bear high similarity. A semi-supervised approach was also presented in [35] for predicting student performance when solving problems and activities in narrative-centered learning environments. A semi-supervised framework was designed in [36] for concept extraction from MOOC textual content. Finally, a SSC model incorporating the Expectation–Maximization (EM) technique was introduced in [37] for mining outcomes and prerequisites within educational textbooks of different disciplines.

5.3.6 Discussion

The first conclusion that can be drawn from the analysis of the aforementioned studies is that SSL results to very accurate predictive learning models harnessing the information hidden in both labeled and unlabeled data. In most studies, SSL algorithms produce very accurate early-warning models for the timely prediction of students at risk of school failure, thus enabling timely support actions and specialized intervention strategies to poor performers with a view to improving their learning performance. These findings open up new horizons and challenges for scientists and researchers concerned.

In addition, it is evident that different SSL methodologies have been applied, most of which incorporate the self-training and co-training concept. However, there is no single self-labeled method that performs best across all datasets. This obviously depends on the dataset used in the research, as well as the features of students which form the set of input attributes. Some of these studies employ input attributes related only to students' grades during their studies. It is worth mentioning that two studies [7, 10] exploit a wide range of attributes, comprising demographic data, academic achievements of learners, and online activity data in the university LMS.

Finally, the review indicates that most of studies deal with the implementation of SSC methods, while SSR has not been given sufficient attention so far. In addition, predicting student performance stands out amongst other output attributes.

All the above discussed studies are summarized in Table 5.1 including information about the structure of the dataset used in each study, the education level, the type of the output attribute, the data mining task, the SSL methods used for constructing the predictive learning model and the evaluation metric employed. The last column concerns the possibility of early prognosis of student learning outcomes.

5.4 The Potential of SSL in the Education Field

Consider the following problem: We have a set

$$L = \{(x_1, y_1), (x_2, y_2), \ldots, (x_k, y_k)\}$$

of k labeled instances corresponding to k students enrolled in a university course for a specified time period, for example in the previous academic year. Each instance

$$x_i = (x_{i1}, x_{i2}, \ldots, x_{in})$$

is a n-dimensional vector of n student features, hereinafter referred to as input attributes, where x_{ij} corresponds to the value of the j-th feature of the i-th student. The output attribute Y refers to the student performance in the final examinations of the course. In this problem, Y is a binary categorical attribute with

$$y_i \in \{pass, fail\}$$

We have also a set

$$U = \{x_1, x_2, \ldots, x_m\}$$

of m unlabeled instances corresponding to m students enrolled in the same course in the current academic year, where each instance x_i is also is a n-dimensional vector of the same features. In this context, we want to accurately predict students' performance (i.e. the value of the output attribute Y) in the forthcoming examinations, which will take place at the end of the academic year.

Table 5.1 Summarizing the discussed SSL studies in the fields of EDM and LA

Paper	#Input features	#Instances	Education level	Output feature and values	Task	SSL methods	Metrics	Early prediction
[20]	15	344	Open university	Unit performance {pass, fail}	Classification	Self-training, Democratic, Co-training, Tri-training, De-tri-training, Rel-Rasco, Rasco	Accuracy, Specificity	Yes
[21]	8	553	Traditional university	Unit performance {pass, fail}	Classification	Tri-training	Accuracy	Yes
[22]	43	1334	Traditional university	Course performance {graduating, studying, first-warned, second-warned, stop studying}	Classification	Self-training	Accuracy	Yes
[23]	43	1334	Traditional university	Course performance {graduating, studying, first-warned, second-warned, stop studying}	Classification	Tri-training	Accuracy	Yes
[24]	43	1334	Traditional university	Course performance {graduating, studying, first-warned, second-warned, stop studying}	Classification	Self-training	Accuracy	Yes
[25]	28	1317	Traditional university	Course performance {graduating, studying, stop studying}	Classification	Co-training	TP-Rate, F-score, AUC	Yes
	28	186						
[7]	50	1073	Open university	Unit performance {pass, fail}	Classification	Co-training	Accuracy, F-score	Yes

(continued)

Table 5.1 (continued)

Paper	#Input features	#Instances	Education level	Output feature and values	Task	SSL methods	Metrics	Early prediction
[27]	15	344	Open university	Dropout {yes, no}	Classification	Self-training, Co-training, De-tri-training, Rasco, Rel-Rasco, Tri-training, Democratic	Accuracy	Yes
[28]	10	340	High school	Grade level {0–9,10–14,15–17,18–20}	Classification	Self-training, Co-training, De-tri-training, Rasco, Tri-training, Democratic	Accuracy	Yes
[29]	10	799	High school	Grade level {0–9,10–14,15–17,18–20}	Classification	Self-training, Co-training, Tri-training	Accuracy	No
[30]	N/S	5388	Traditional university	Grade level {I, II, III, IV}	Classification	Label Propagation	Precision	No
[10]	90	1073	Open university	Examination grade [0, 10]	Regression	MSSRA	MAE, RAE, RMSE, PCC	Yes
[31]	8	15,362	Secondary education	Examination grade N/S	Regression	Semi-Supervised Multi-Task Regression	MSE	No
[32]	26	1073	Open university	Examination grade [0, 10]	Regression	Coreg	MAE	Yes

This problem could be addressed with the use of traditional supervised methods as follows:

- Train a classification algorithm c on L for building a classification model h, which is subsequently applied in U for predicting the performance of current students. This approach generally leads to very efficient learning models, since all available information about students is utilized to create the model, but at the end of the academic year (Fig. 5.2). Hence, the prediction time point leaves no room for the timely identification of weak students and providing support to them. Moreover, it is rather non-pedagogical to inform a student that he/she is very likely to fail just before the examinations.
- Train a classification algorithm c on a subset $L_i \subseteq L$, by limiting the available information about students to a particular subperiod $[0, t_i]$, for example the first semester of the academic year (Fig. 5.3). After that, the produced classification model h is applied in U_i for predicting student performance. In this way, it is possible to predict timely poor performers and provide effective interventions. However, the available data of L and U are not yet fully exploited.

It should be noted that the unlabeled dataset U is not used during the training process in both cases. Generally, three main shortcomings arising from these scenarios:

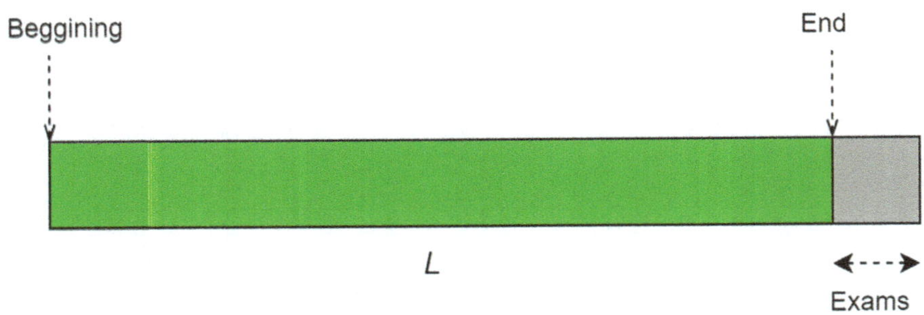

Fig. 5.2 Using supervised learning for predicting student performance (1st scenario)

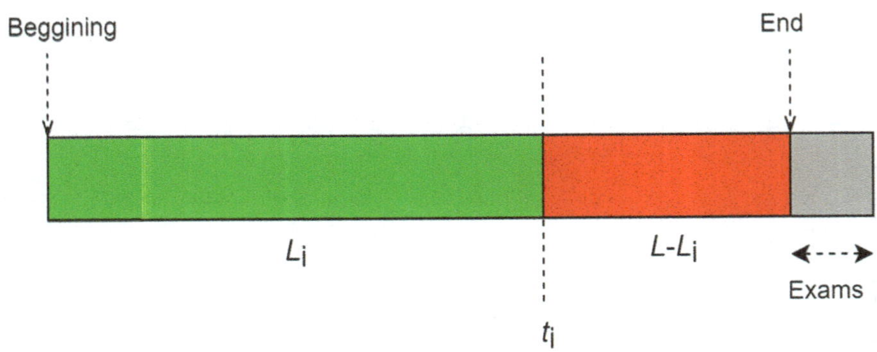

Fig. 5.3 Using supervised learning for predicting student performance (2nd scenario)

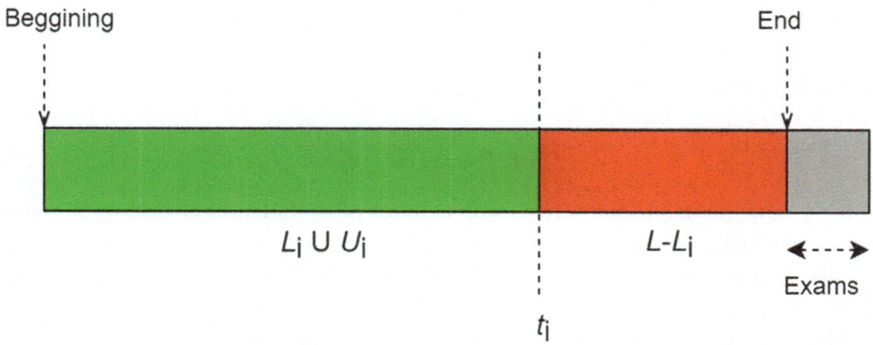

Fig. 5.4 Using semi-supervised learning for predicting student performance

1. A sufficient amount of labeled data is required for training the supervised algorithm c.
2. Traditional supervised approaches commonly build very efficient learning models utilizing all available information about students, but at the end of the academic year. Unfortunately, these approaches do not emphasize to the early prognosis aspect of the problem.
3. The unlabeled dataset U is not used.

Semi-Supervised Learning is the appropriate approach for exploiting both datasets L and U and improving the learning performance [8]. To this end, a supervised classification algorithm c is initially trained on L_i and subsequently is applied on U_i for classifying the unlabeled data (Fig. 5.4).

The most confident predictions are integrated into L_i and c is retrained through an iterative learning process. This approach utilizes all available information regarding students of the previous academic year and the current one. The learning model h_i developed in this way boosts the predictive performance in most cases, demonstrating the potential of SSL for both accurate and timely prediction of student learning outcomes. In some other cases, h achieves the same performance as supervised models, but with fewer labeled examples, thereby reducing the training cost [8].

In this way, a dataset

$$L = \{x_i, y_i\}_{i=1}^{k}$$

consisting of the values

$$\{x_i\}_{i=1}^{k} = \{(x_{i1}, x_{i2}, \ldots, x_{in})\}_{i=1}^{k}$$

of the k input attributes $X_1, X_2, \ldots, X_n$, regarding features of k students enrolled in a unit course in the previous academic year or a semester, and the value y_i of the output attribute Y (i.e. the learning outcome) could form the labeled dataset. In addition, a dataset

$$U = \{x_i\}_{i=1}^{m}$$

consisting of the values

$$\{(x_{i1}, x_{i2}, \ldots, x_{in})\}_{i=1}^{m}$$

of the k same input attributes, regarding features of m students enrolled in the same unit course in the current academic year or semester could form the unlabeled dataset. Finally, a supervised algorithm is trained on $L \cup U$ for building a classification or a regression model h.

The proposed approach can efficiently handle the temporal perspective of a predictive problem, thus allowing timely support and effective interventions to failure-prone students. In this context, the produced model may serve as an early warning system. What we need to do is to select a time-period $[0, t_i]$ which results to a very accurate predictive model and apply it to predict student learning outcomes. Moreover, we can determine the effectiveness of support strategies for specific groups of identified weak students by applying the predictive model to the next time period $[t_i, t_{i+1}]$.

In cases that a course does not involve the same input attributes from one academic year to another or from one learning period to another, transfer learning [38] could assist as for exploiting the knowledge retrieved from a unit course in one period for improving the predictive performance of a learning model to the next one in combination with a self-labeled method.

5.5 Conclusions

Semi-Supervised Learning has received considerable attention among machine learning researchers over the last few years blending the concepts of supervised and unsupervised paradigms. As a result, a plethora of SSL methodologies and algorithms have been introduced and employed with great success for solving a variety of machine learning problems in many scientific fields, among which the education field as well. In the present study, an effort was made to expand current research and provide a comprehensive review of SSL applications in the area of EDM and LA. To this end, a plethora of notable studies were discussed and classified according to the predicted learning outcome in each of them.

Overall, the review highlights the potential of the SSL approach in the education field too for building highly efficient predictive models. What is more important is that these models may serve as an early alert system for the timely prognosis of students at risk of failure, enabling appropriate assistance and intervention strategies to low performers. The experimental results in most studies demonstrated the superiority of SSL methods over representative supervised methods for extracting useful knowledge from both labeled and unlabeled data, resolving important educational and supporting decision making processes.

For future work we consider the issue of building interpretable and explainable learning models in the education field. To put it simply, we aim to understand the predictions of a machine learning model when predicting student learning outcomes (e.g., why a student failed a final exam) and extract more valuable information [39].

References

1. F. Mikre, The roles of information communication technologies in education: review article with emphasis to the computer and internet. Ethiop. J. Educ. Sci. **6**(2), 109–126 (2011)
2. C. Romero, S. Ventura, Educational data mining: a review of the state of the art. IEEE Trans. Syst. Man, Cybern. Part C (Applications Rev.) **40**(6), 601–618 (2010)
3. C. Romero, S. Ventura, Data mining in education. RWiley Interdiscip. Rev. Data Min. Knowl. Discov. **3**(1), 12–27 (2013)
4. G. Siemens, R.S.J. d Baker, Learning analytics and educational data mining: towards communication and collaboration, in *Proceedings of the 2nd International Conference on Learning Analytics and Knowledge* (2012), pp. 252–254
5. C. Romero, S. Ventura, Educational data mining and learning analytics: an updated survey. Wiley Interdiscip. Rev. Data Min. Knowl. Discov. **10**(3), e1355 (2020)
6. Z.-H. Zhou, A brief introduction to weakly supervised learning. Natl. Sci. Rev. **5**(1), 44–53 (2018)
7. G. Kostopoulos, S. Karlos, S. Kotsiantis, Multiview learning for early prognosis of academic performance: a case study. IEEE Trans. Learn. Technol. **12**(2) (2019)
8. X. Zhu, A.B. Goldberg, Introduction to semi-supervised learning. Synth. Lect. Artif. Intell. Mach. Learn. **3**(1), 1–130 (2009)
9. O. Chapelle, B. Scholkopf, A. Zien, Semi-supervised learning (O. Chapelle, et al., eds. 2006) [book reviews]. IEEE Trans. Neural Netw. **20**(3), 542 (2009)
10. G. Kostopoulos, S. Kotsiantis, N. Fazakis, G. Koutsonikos, C. Pierrakeas, A semi-supervised regression algorithm for grade prediction of students in distance learning courses. Int. J. Artif. Intell. Tools **28**(4) (2019)
11. A.B. Goldberg, X. Zhu, New directions in semi-supervised learning, University of Wisconsin–Madison (2010)
12. D. Yarowsky, Unsupervised word sense disambiguation rivaling supervised methods, in *Proceedings of the 33rd Annual Meeting on Association for Computational Linguistics* (1995), pp. 189–196
13. A. Blum, T. Mitchell, Combining labeled and unlabeled data with co-training, in *Proceedings of the Eleventh Annual Conference on Computational Learning Theory* (1998), pp. 92–100
14. V. Ng, C. Cardie, Weakly supervised natural language learning without redundant views, in *Proceedings of the 2003 Human Language Technology Conference of the North American Chapter of the Association for Computational Linguistics* (2003)
15. J. Tanha, M. van Someren, H. Afsarmanesh, Semi-supervised self-training for decision tree classifiers. Int. J. Mach. Learn. Cybern. **8**(1), 355–370 (2017)
16. Z.-H. Zhou, M. Li, Tri-training: exploiting unlabeled data using three classifiers. IEEE Trans. Knowl. Data Eng. **17**(11), 1529–1541 (2005)
17. C. Deng, M.Z. Guo, A new co-training-style random forest for computer aided diagnosis. J. Intell. Inf. Syst. **36**(3), 253–281 (2011)
18. Y. Yaslan, Z. Cataltepe, Co-training with relevant random subspaces. Neurocomputing **73**(10–12), 1652–1661 (2010)
19. P.M. Moreno-Marcos, C. Alario-Hoyos, P.J. Muñoz-Merino, C.D. Kloos, Prediction in MOOCs: a review and future research directions. IEEE Trans. Learn. Technol. (2018)

20. G. Kostopoulos, S. Kotsiantis, P. Pintelas, Predicting student performance in distance higher education using semi-supervised techniques, in *Model and Data Engineering* (Springer, 2015), pp. 259–270
21. V.T.N. Chau, N.H. Phung, Combining self-training and tri-training for course-level student classification, in *2018 International Conference on Engineering, Applied Sciences, and Technology (ICEAST)* (2018), pp. 1–4
22. V.T.N. Chau, N.H. Phung, A random forest-based self-training algorithm for study status prediction at the program level: minSemi-RF, in *International Workshop on Multi-disciplinary Trends in Artificial Intelligence* (2016), pp. 219–230
23. V.T.N. Chau, N.H. Phung, A robust random forest-based tri-training algorithm for early in-trouble student prediction, in *2017 4th NAFOSTED Conference on Information and Computer Science* (2017), pp. 84–89
24. V.T.N. Chau, N.H. Phung, On Semi-supervised learning with sparse data handling for educational data classification, in *International Conference on Future Data and Security Engineering* (2017), pp. 154–167
25. N.D. Hoang, V.T.N. Chau, N.H. Phung, Combining transfer learning and co-training for student classification in an academic credit system, in *2016 IEEE RIVF International Conference on Computing & Communication Technologies, Research, Innovation, and Vision for the Future (RIVF)* (2016), pp. 55–60
26. S. Utami, I. Winarni, S.K. Handayani, F.R. Zuhairi, When and who dropouts from distance education? Turkish Online J. Distance Educ. **21**(2), 141–152 (2020)
27. G. Kostopoulos, S. Kotsiantis, P. Pintelas, Estimating student dropout in distance higher education using semi-supervised techniques, in *ACM International Conference Proceeding Series*, vol. 01–03 Oct (2015)
28. G. Kostopoulos, I.E. Livieris, S. Kotsiantis, V. Tampakas, Enhancing high school students' performance based on semi-supervised methods, in *2017 8th International Conference on Information, Intelligence, Systems & Applications (IISA)* (2017), pp. 1–6
29. I.E. Livieris, K. Drakopoulou, T.A. Mikropoulos, V. Tampakas, P. Pintelas, An ensemble-based semi-supervised approach for predicting students' performance, in *Research on e-Learning and ICT in Education* (Springer, 2018), pp. 25–42
30. H. Yao, M. Nie, H. Su, H. Xia, D. Lian, Predicting academic performance via semi-supervised learning with constructed campus social network, in *International Conference on Database Systems for Advanced Applications* (2017), pp. 597–609
31. Y. Zhang, D.-Y. Yeung, Semi-supervised multi-task regression, in *Joint European Conference on Machine Learning and Knowledge Discovery in Databases* (2009), pp. 617–631
32. S. Karlos, G. Kostopoulos, S. Kotsiantis, Predicting and interpreting students' grades in distance higher education through a semi-regression method. Appl. Sci. **10**(23), 8413 (2020)
33. Z.-H. Zhou, M. Li, Semi-supervised regression with co-training. IJCAI **5**, 908–913 (2005)
34. S. Jing, O.C. Santos, J.G. Boticario, C. Romero, M. Pechenizkiy, A. Merceron, Automatic grading of short answers for MOOC via semi-supervised document clustering, in *EDM* (2015), pp. 554–555
35. W. Min, B.W. Mott, J.P. Rowe, J.C. Lester, Leveraging semi-supervised learning to predict student problem-solving performance in narrative-centered learning environments, in *International Conference on Intelligent Tutoring Systems* (2014), pp. 664–665
36. Z. Jiang, Y. Zhang, X. Li, Moocon: a framework for semi-supervised concept extraction from Mooc content, in *International Conference on Database Systems for Advanced Applications* (2017), pp. 303–315
37. I. Labutov, Y. Huang, P. Brusilovsky, D. He, Semi-supervised techniques for mining learning outcomes and prerequisites, in *Proceedings of the 23rd ACM SIGKDD International Conference on Knowledge Discovery and Data Mining* (2017), pp. 907–915
38. M. Tsiakmaki, G. Kostopoulos, S. Kotsiantis, O. Ragos, Transfer learning from deep neural networks for predicting student performance, *Appl. Sci.* **10**(6) (2020)
39. Z.C. Lipton, The mythos of model interpretability. Queue **16**(3), 31–57 (2018)

Part III
Machine Learning/Deep Learning in Security

Chapter 6
Survey of Machine Learning Approaches in Radiation Data Analytics Pertained to Nuclear Security

Miltiadis Alamaniotis and Alexander Heifetz

Abstract The increasing concerns over the use of nuclear materials for malevolent purposes (i.e., terrorist attacks) have fueled the interest in developing technologies that can detect hidden nuclear material before its use. The process of detecting and identifying nuclear materials for non-reported purposes is under the umbrella of nuclear security. Among several areas that contribute to the security and safeguards of nuclear materials, radiation data analytics has recently been marked as an area of high potential. Therefore, there is an increasing trend in applying machine learning for developing data analysis methods applied to radiation signals aiming at identifying patterns of interest associated with nuclear materials. This chapter aspires in providing a comprehensive survey and discussion of machine learning and data analytics methods pertained to nuclear security. The chapter will also discuss further trends and how data analytics can further enhance nuclear security by effectively analyzing radiation data.

Keywords Machine learning · Data analytics · Nuclear security · Radiation data

6.1 Introduction

The historic terrorist attack in New York on 9/11 was not only extensively disastrous with thousands of losses, but also was the turning point for revising the security architecture of our societies and identifying gaps and vulnerabilities in it [1]. Among the scenarios that were examined, the most frightening ones involved the conduction of terroristic attacks by the use of nuclear materials [2]. The consequences of such

M. Alamaniotis (✉)
Department of Electrical and Computer Engineering, The University of Texas at San Antonio, 1 UTSA Cir., San Antonio, TX 78249, USA
e-mail: miltos.alamaniotis@utsa.edu

A. Heifetz
Nuclear Science and Engineering Division, Argonne National Laboratory, Argonne, IL 60439, USA
e-mail: aheifetz@anl.gov

G. A. Tsihrintzis et al. (eds.), *Advances in Machine Learning/Deep Learning-based Technologies*, Learning and Analytics in Intelligent Systems 23,
https://doi.org/10.1007/978-3-030-76794-5_6

attacks have been identified as extremely severe while having a long-term negative impact on the feeling of security in society [3]. Therefore, deterrence of nuclear terrorism is of paramount importance that may be attained by developing a defense-in-depth architecture for nuclear security.

In this nuclear security architecture, the safeguard of nuclear materials plays an important role. By fully securing the accountability of nuclear material, we minimize the risk of having the material stolen and being used for nefarious purposes. Therefore, one of the first lines of defense against nuclear terrorism is monitoring the use, storage, and movement of nuclear material in designated facilities, and surveying urban areas to detect hidden radioactive sources [4]. Up to now, the implementation of this line of defense is based on the measurement and detection of radiation [5]. The development of radiation sensor technologies is a complex process that requires the involvement of several domains from science and engineering. One of the preeminent domains in developing efficient and accurate sensing technologies is the domain of data analytics [6], whose main goal is to analyze the obtained radiation data and make inferences about the presence of a threat or not among their constituents [7].

Advancements in machine learning (ML) in the last decade accommodated the fast and efficient processing of vast amounts of data acquired from various—often heterogeneous—sensors and launched the era of *intelligent data analysis* [8]. Notably, machine learning has significantly boosted the use of data analytics in a high number of application domains and paved the way for developing automated and autonomous systems. Inevitably, the field of nuclear security was also influenced by machine learning, with several methods been developed or are under development. Though these methods have different aims, their ultimate goal is to support the analysis of data to identify patterns of interest that are correlated with the use of nuclear materials [9].

Radiation data of interest to nuclear security take two forms and more specifically: (i) energy spectrum, that is a histogram expressing the number of counts (i.e., events) per energy interval, and (ii) count rate signal that is a time series denoting the total number of counts detected within a specific time interval. Figure 6.1 provides an

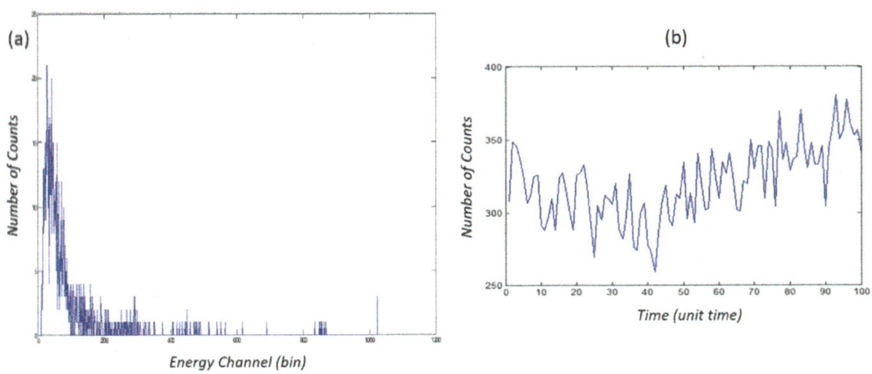

Fig. 6.1 Example of radiation data: **a** energy spectrum, **b** count rate signal (time series)

example of two measured radiation datasets, namely, a spectrum and a count rate signal. The contents of the data are the result of the contribution of various radioactive sources by convoluting the individual contributions [5]. Therefore, data analysis aims at identifying and characterizing the contents of the data that will subsequently lead to inferences over the detection of a threat or not. Detection of threat is based on identifying patterns that belong to known radionuclides and uniquely characterize them (known as signature patterns) [10]. However, their detection imposes significant challenges due to a course of factors that include, the stochasticity of radioactive processes, the detector-source distance, the ambient environment, the intentional shielding of the sources, and the instrument's (electronic) noise [5].

Machine learning has offered a variety of tools that serve the purposes of nuclear security applications. Of importance is the inherent capability of those tools to learn the signatures, and then seek for them in acquired datasets. Except for identifying signatures, learning tools may be used for other purposes such as estimating background radiation patterns, detecting anomalies, characterizing properties of known sources, optimal sensor placement, and data fusion, as presented in Fig. 6.2. The list of machine learning applications in nuclear security will only grow thicker being boosted by the advances in computing science that allow the processing of increasing volumes of data in a fast and accurate way [11].

This chapter focuses on presenting a variety of machine learning methods that have been applied in nuclear security applications. Its goal is twofold, and more

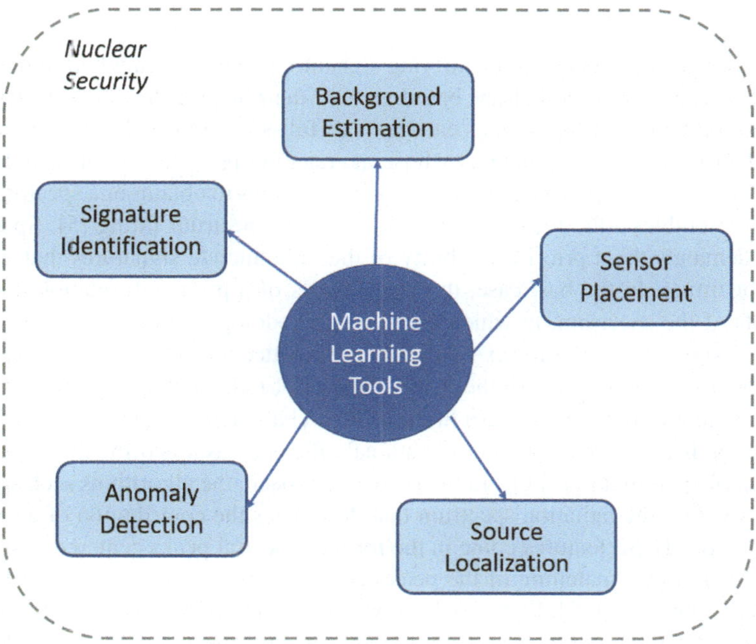

Fig. 6.2 Machine learning applications in nuclear security

particularly, to a provide comprehensive survey of the existing methods, and become a point of inspiration for future research directions. To that end, the rest of the chapter is organized as follows: each section presents an area of nuclear security depicted in Fig. 6.2 followed by a brief description of the machine learning methodologies developed for this area. It should be noted that the chapter strictly focuses on methodologies that perform learning and does not discuss any other type methods from artificial intelligence that do incorporate some type of learning [12].

6.2 Machine Learning Methodologies in Nuclear Security

The application of data analytics in nuclear security requests the development of methodologies that have as ground one or more of machine learning tools. Notably, the methodologies developed require more than just selecting a learning tool; determination of input and output structure, knowledge representation, data reduction, availability and reliability of data, determination of training and testing datasets are also some of the issues that the modeler should take into consideration in developing a data analysis methodology.

6.2.1 Nuclear Signature Identification

Methodologies developed for identifying signature patterns are applied to the analysis of radiation spectra and are based on two main approaches [13]. In the first one, the spectrum is handled in its entirety (i.e., full-spectrum analysis) and the goal is to identify the set of signatures whose aggregation provides the observed spectrum. The process of putting together nuclear signatures to obtain–or a spectrum that closely resembles—the measured one is known as spectrum fitting [5]. Spectrum fitting demands the a priori availability of the radionuclide signatures that will be used for fitting. In the best case, there are pieces of a priori information over the contents of the spectrum, in which case the methodology utilizes a specific population of signatures—the lower the number the better for getting a more accurate result, while in the worst case the whole number of radionuclide signatures (above 700) is required. In the latter case fitting is high challenging—almost impossible—concerning high accuracy and computational efficiency as it is difficult to check all the possible permutations [14]. In the second approach, the algorithms seek specific features within the radiation spectrum that designates the contribution of a specific radionuclide. These features come in the form of spectral peaks centered at specific energy that allows matching of the peak center energy to peak energies of known radioactive nuclides [15]. Figure 6.3 shows an example of the spectrum fitting and peak identification approaches, while the presence of noise in the form of Poisson fluctuation is apparent.

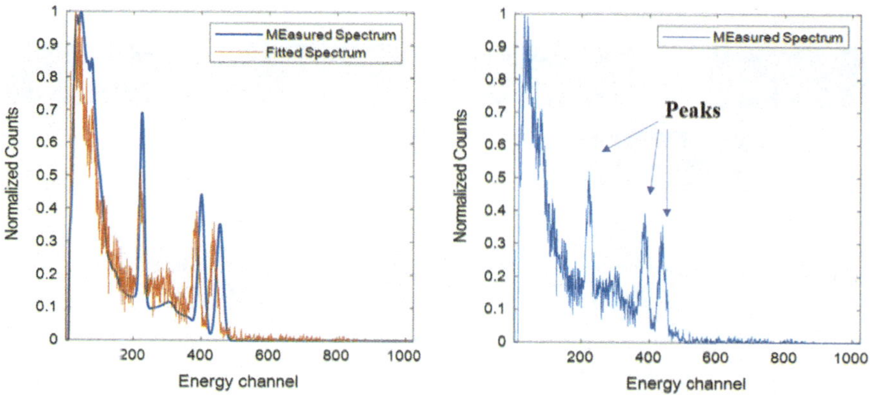

Fig. 6.3 Example of **a** energy spectrum fitting, and **b** peak identification

Machine learning methodologies have been developed implementing with each one having its advantages and disadvantages. Intelligent data analytics research is highly active in this area mainly driven by the criticality of the nuclear security mission and the need to overcome the computational challenges. It should be noted that the proposed methodologies either adopt exclusively tools from machine learning or adopt a combination of machine learning with other analysis tools from the broader area of artificial intelligence and statistics [16].

6.2.1.1 Spectrum Analysis

Machine learning has been widely used in analyzing nuclear spectra by using full-spectrum analysis. There is a variety of machine learning tools that have been used either as standalone tools or in combination with other tools. In the latter case, significance is the use of preprocessing tools that identify features of the full spectrum and forward them to the ML tool. Figure 6.4 provides the general block diagrams of the methodologies followed so that a full spectrum may be analyzed using machine learning: there are two parts, namely, the training and the analysis.

The training part (Fig. 6.4a) exploits the full spectrum in two ways; (i) the full spectrum is utilized for training the ML tool, or (ii) extracting features and then provide them to the tool. In both cases, a training dataset is created that is used for training the ML tools that have been selected. If preprocessing is needed, the features that can be used to uniquely identify radionuclide patterns are selected. For instance, the number of counts with a specific region of interest (ROI) in the spectrum or the number of peaks in the whole spectrum may be informative features. The type of training performed is supervised given that for known input spectra, the label of the nuclide is known. Once the tool has been trained, then it is suitable for identifying for spectrum analysis.

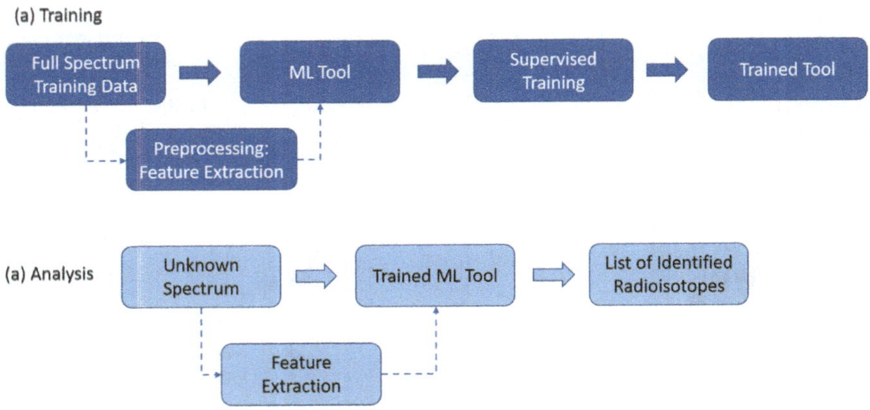

Fig. 6.4 Machine learning steps in implementing full-spectrum analysis **a** tool training, and **b** analysis of spectrum

Notably, the use of training data is essential for building effective detection systems. In nuclear security applications, systems, which make use of gamma-ray spectra, for instance, populate their training datasets with spectra that have been either simulated or experimentally taken. It should be mentioned at this point that the use of features as an input to the ML tools is preferable as compared to the use of the whole spectrum given that the training time will be considerably lower.

The second part (Fig. 6.4b), which contains the spectrum analysis, makes use of the pre-trained ML tool to identify the constituents of the spectrum. In particular, the unknown spectrum is provided as an input to the ML tool that processes it and outputs the list of identified radionuclide patterns. In the case that the ML tool has been trained on a set of features, then a feature extraction step precedes as shown in Fig. 6.4b.

Based on the above approaches there is a wide set of machine learning tools that have been used for spectrum processing in nuclear security. The general framework of these tools is given below.

Artificial Neural Networks: One of the most common ML tools in information processing is the artificial neural networks (ANN) that have been applied to a variety of problems in various domains. ANNs have also found use in full-spectrum processing of radiation signals with all proposed ANN share a common framework that is presented in Fig. 6.5. In particular, the full spectrum is provided as the input to the neural network, which is trained on a specific set of radionuclides and gives as an output the list of identified patterns.

In the scientific literature, there is a variety of works that have been developed adopting artificial neural networks. Notably, the proposed approaches utilize various types of ANN to analyze the radiation spectrum. A simple feedforward neural network approach is proposed in [17–20] aiming at identifying full signature patterns in the radiation spectra. Furthermore, the synergism of feedforward ANN

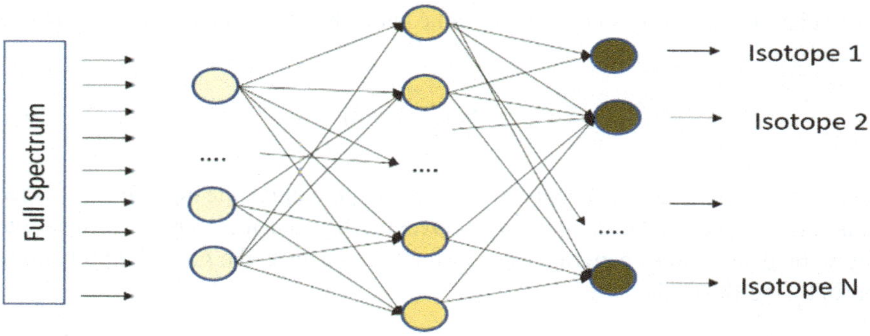

Fig. 6.5 Artificial Neural Network full spectrum analysis

with Karhunen—Loeve transform is presented in [21], while a method that utilizes synergistically ANN with Bayes rules is proposed in [22]. Moreover, the use of more sophisticated ANN tools was introduced in [23, 24], where convolutional neural network (CNN) models were adopted for radionuclide identification.

Multiple Linear Regression: One of the first tools in analyzing radiation spectra is the well-known multiple linear regression (MLR). MLR assumes that the measured spectrum may be considered as the linear aggregation of the signature patterns, where each signature has its contribution. In particular:

$$M = a_1 S_1 + a_2 S_2 + \cdots + a_N S_N \tag{6.1}$$

where M is the measured spectrum, $S_\#$ is the nuclide signature spectrum, $\alpha_\#$ is the contribution of the signature in the measurement, and N is the population of the nuclides. In the ideal case, the N is equal to the number of known nuclides—over 700—with the linear coefficients being equal to zero for those nuclides with no contributions in the measurement. The extremely high complexity of the ideal case requires a very high volume of computational resources, thus, making it impossible to be used (especially in real-time applications). In practice, a selection process is required to precede the MLR analysis that designates a subset of the radionuclides that may be used in analysis either automatic or manual [25, 26].

Throughout the years, MLR has been proposed in various approaches for radiation data analytics. In [27], MLR was firstly proposed as a standalone tool, while in [28] was combined with fuzzy logic [28]. Other works proposed the use of MLR with a correlation measure to identify nuclide patterns as discussed in [29], while the use of MLR together with chi-square-based objectives in radionuclide detection was introduced in [30].

Bayesian Inference: Data analytics methods for analyzing radiation data using Bayesian linear regression have also been proposed. As opposed to MLR, Bayesian regression is formed by using probability distributions instead of point estimates. In particular, the response is not a single value but is drawn from a normal distribution given by:

$$y \sim N(\alpha^{\mathrm{T}}\mathrm{X}, \sigma^2\mathrm{I}) \tag{6.2}$$

where α are the linear coefficients, X is the vector containing the input data, σ^2 is the data variance and I is the identity matrix. Overall, Bayesian methods utilize spectrum fitting by using library patterns, whose values are taken as the posterior probability given by the Bayes theorem:

$$P(A|B) = \frac{P(A|B) \cdot P(A)}{P(B)} \tag{6.3}$$

where $P()$ denotes a probability with $P(A|B)$ being the conditional probability of the event A given that B occurs. Thus, model parameters assume the use of prior probabilities over the contribution of each nuclide in the spectrum (usually equal probability if no information is available as shown in Fig. 6.6). In radiation data analytics, $P(A)$ and $P(B)$ are marked as the prior probabilities of the radionuclide signature pattern contributions (i.e., the α parameters in the linear regression model) and are used to implement Bayesian regression where each regressor coincides with the radiation pattern of a nuclide. Utilization of Bayesian approaches allows the determination of the model complexity (i.e., the number of nuclides in the regression model) in an automated way as opposed to MLR where a manual selection is required.

One of the main assumptions made in using Bayesian inference approaches in nuclear security is that the set of radionuclides we may find in the measured spectrum is limited to a small number (e.g., up to 50). This allows to consider the prior probabilities to be uniformly distributed (i.e., of equal value) or identifying rough estimates of them. This limited set of nuclides presumes the availability of prior knowledge over the set of nuclides that we may encounter in our nuclear search [31].

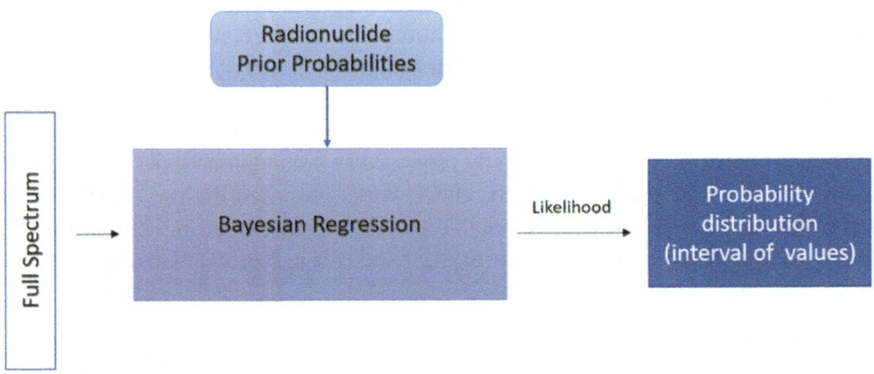

Fig. 6.6 General frame of using Bayesian Regression in full spectrum analysis

Similar to MLR, ideally all the available radionuclides must be considered when analyzing the full spectrum, and therefore, the computational and time constraints are retained.

Bayesian approaches in radiation data analytics in a machine learning framework have been proposed in various forms. In [32], an identification algorithm has been proposed utilizing sequential Bayesian Learning, while in [33] a simple Bayesian regression algorithm for the identification of weak signatures is presented and tested. Furthermore, a simple Bayesian inference approach, which was tested on a set of 6 nuclides, was introduced in [34]. Lastly, an approach utilizing Bayesian inference networks in nuclide identification was introduced in [35].

6.2.2 *Background Radiation Estimation*

One of the main concerns in nuclear security is the modeling of background radiation, especially in hidden source search applications. Background contribution is inevitably present in a radiation measurement given that radiation emission from naturally occurring radioactive materials (NORM) is omnipresent. Though background radiation may be defined as the radiation that emerged from NORM, in a nuclear security framework background may be considered the contribution that comes from radioactive sources that are not identified as a threat [36]. For instance, medical radioisotopes, which are used for treatment purposes, may be considered as part of the background radiation in a hospital unit.

Therefore, it is essential in knowing the background contribution either in the number of counts—in the case of time series data—or the form of an energy distribution pattern in the case of a spectrum. Given the random nature of the background and the dynamically varying conditions, getting the exact number of background counts is highly challenging. The degree of challenge only increases in nuclear search of hidden radioactive material with a mobile detector—as is the case in nuclear searches—because the contents of the ambient environment may be unknown. Notably, in spectrum analysis, the background spectrum may mask weak signatures coming from radioactive sources of interest (i.e., threats) given that a radiation measurement is the aggregation of the individual signatures [5].

Background estimation has profound significance in nuclear security to accommodate data processing, as shown in Fig. 6.7. In particular, its accurate estimation allows its subtraction from the measurement, hence, providing a "clean" signal where the radioactive source patterns are easily recognizable.

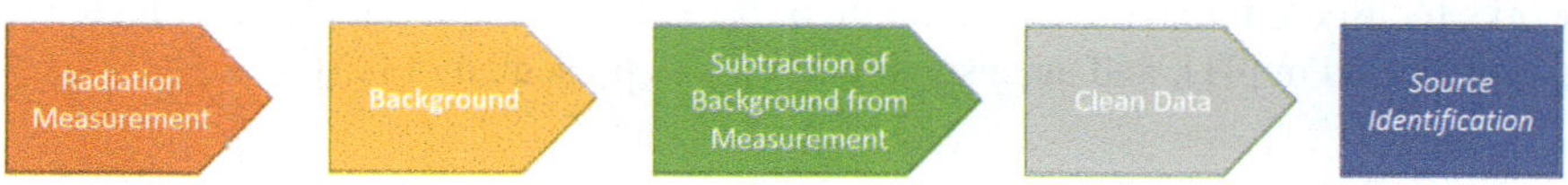

Fig. 6.7 Block diagram of data processing utilizing background data in nuclear security

The most common approach in estimating the background contribution is by obtaining a measurement in a controlled environment—for example, a laboratory—where there is no radioactive material identified as a threat. However, a measurement only provides an instance of the background and does not fully fulfill the needs of nuclear security applications especially in cases where the radiation environment dynamically changes. To that end, machine learning has been utilized to provide effective solutions by estimating the contribution of radiation background in measured data. There is a limited variety of ML tools that have been proposed for background estimation and modeling as given below.

Gaussian Processes: Kernel machines are a popular class of learning algorithms with wide use in various applications. One of the preeminent learning tools that belong to the class of kernel machines is the tool of Gaussian Processes (GP). GP is a Bayesian inference model that forms a predictive distribution that follows a normal distribution. The mean and the covariance functions of GPs are modeled as a function of a kernel, which takes the following form:

$$k(x_1, x_2) = f(x_1)^T f(x_2) \tag{6.4}$$

where *f*() is the so-called basis function. The framework of GPs provides a predictive distribution that is Gaussian and its parameters, i.e., mean and variance, are given by:

$$m(x_{N+1}) = \boldsymbol{k}^T \boldsymbol{C}_N^{-1} \boldsymbol{t}_N \tag{6.5}$$

$$\sigma^2(x_{N+1}) = k - \boldsymbol{k}^T \boldsymbol{C}_N^{-1} \boldsymbol{k} \tag{6.6}$$

with N, C_N, and k being the number, the covariance matrix, and kernel evaluation of training data respectively, while $\boldsymbol{t}_N$ is the vector of target data of the training set, and $\mathbf{x}_{N+1}$ is the incoming unknown data point. Full details of the GP framework may be found in [37], while the general framework for making background estimation is given in Fig. 6.8.

GP have been the main constituent in a variety of methodologies that have been proposed for background estimation either in the form of time series or spectrum data. In particular, in [38] a methodology based on GP is used to estimate the background spectrum in low count spectra with no prior information available. In a similar setting, an ensemble of GP was used in [39] to estimate the background spectrum, while in [40] two GPs were used to capture the curvature of the background signal. Other works focused on time series prediction of background count rate using a single GP as introduced in [41], and an ensemble of GPs as presented in [42].

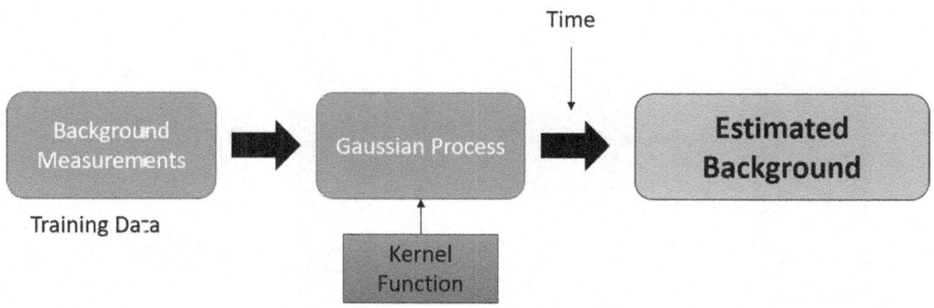

Fig. 6.8 General Framework for background contribution prediction at time t using GP

Statistical Learning Estimation: The domain of statistical methods has also provided tools that have been adopted in radiation data analysis. Several of the statistical methods comprise the foundation upon which sophisticated machine learning methods were built and therefore they are also considered as part of the ML realm.

There is a set of statistical tools that have been proposed for estimating background contribution in radiation spectra. The general framework is very similar to that of the GP: a set of training data is used to fit a model which utilizes the available data to predict the background. In general, the count rate is assumed to follow a probabilistic distribution, usually Poisson or Gaussian, that is subsequently used to form the well-known form of MLE [6].

To that end, maximum likelihood estimation (MLE) has been proposed in [43] to individually estimate the count rate in every channel of an airborne gamma radiation detector. Furthermore, a methodology that initially uses data learning to estimate the Poisson process parameters, and subsequently applies the Poisson model to estimating background counts in spectral peaks was introduced in [44].

6.2.3 *Radiation Sensor Placement*

One of the main concerns in nuclear security is the enhancement of detection probability of radioactive sources that are characterized as threats. The quality of acquired data also depends on the physical constraints imposed in the radiation sensors. One of the constraints has to do with the sensor-source distance that impacts the intensity of source signal signatures in the acquired measurement. The signal emitted from the source degrades as a function of the inverse square of the distance—$1/s^2$—and as a result, the detection probability significantly diminishes some meters away from the source.

Notably, one of the main challenges in nuclear security is the deployment of radiation sensor networks in a way that ensures the detectability of critical radionuclides. The requirement is two-fold: the sensor network must have full coverage of the designated area—to ensure detectability—with the optimal number (i.e.,

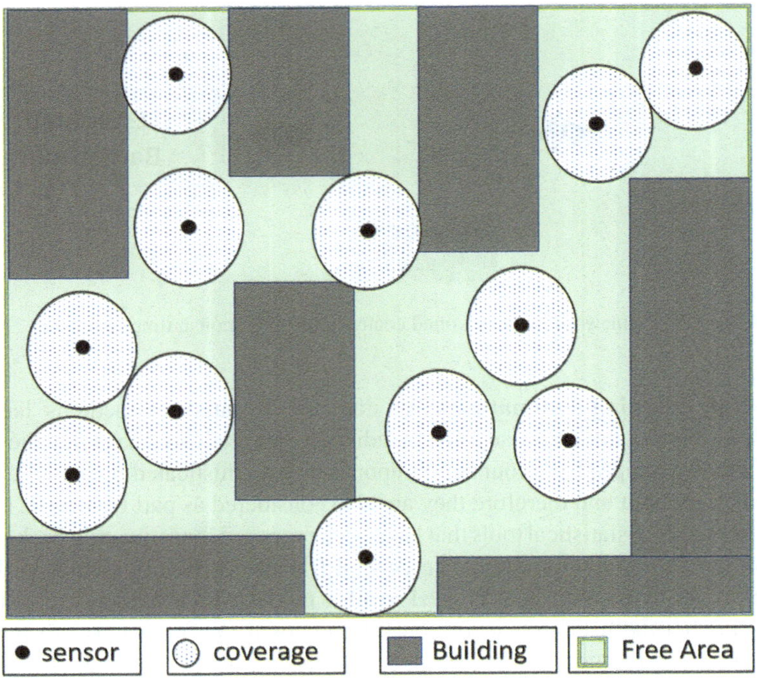

Fig. 6.9 Visualization of radiation sensor network deployment

minimum number) of sensor nodes—to ensure low cost and computational inexpensive operation- as shown in Fig. 6.9. Therefore, data analytics methods have been utilized for identifying the optimal network configuration in the form of sensor placement.

In this frame, the goal of machine learning is to identify those locations on which the radiation sensors should be located to provide the maximum probability of detection. The detection probability is mainly expressed through the maximum coverage of each of the sensors and their minimum overlap. However, it should be noted that in radiation sensors the coverage is a function of many uncontrolled parameters: source-sensor distance, the strength of the source (or amount), and the type of the source. For instance, the same amount of the nuclides U-238 and Cs-137 is detected in different distance lengths, thus, requiring a different number of sensors to enhance detection probability. Therefore, the particularity of radiation sensors is that they have to be deployed based on the source of interest, making the use of analytics essential in making decisions over the sensor placement.

To that end, machine learning has been utilized to offer solutions that accommodate the tradeoff of high detection and a minimum number of sensors. Also, the advantage of machine learning is that it allows the concurrent consideration of the coverage of several sources including background. Sensor placement in nuclear security is still an active area of research with machine learning only recently being adopted with the main ML method proposed in this frame adopting the Bayes theorem as presented below.

Sequential Bayesian Approach: In this approach, the Bayesian formula framework, i.e., Eq. (6.3), is utilized to place a set of mobile sensors in an optimal assembly. The Bayes formula is used sequentially to process a set of experimentally obtained measurements; at every iteration, a single measurement is used and the updated probability is used as the prior probability for the next iteration. Furthermore, a mutual information formula is also adopted as a metric to identify the set of positions (among a set of predetermined discrete positions) that will provide the highest amount of information about the source localization parameters. The proposed approach is fully described and tested in [45].

6.2.4 Source Localization

As mentioned earlier, one of the main goals of nuclear security is the localization of hidden sources in urban environments. It should be noted that the presence of buildings and other materials challenge the identification of the exact location of the radioactive source. Though localization is problem-dependent, however, machine learning offers tools that may be used to localize the source by identifying its parameters independently of the problem at hand. This is still an active research area and there is only limited work done up to this point.

Bayesian Metropolis Approach: In this approach, the localization problem is formulated as a Bayesian parameter estimation problem. Based on the Bayesian formulation a Markov Chain Monte Carlo (MCMC) method is employed to provide a posterior probability density function. Due to computational difficulties, the MCMC is realized using the Metropolis–Hastings algorithm that assumes a candidate density function and draws samples from it. The proposed methodology is described in [46].

Bayesian Inference: In [47], the Bayesian formulation together with adjoint radiation models was used for illicit source localization. The proposed formulation allows for the incorporation of all the previously acquired information in the form of prior probabilities, thus enhancing the detection of the exact location of the source.

Maximum Likelihood Estimation: MLE has also been used for identifying the location of a source by estimating the counts coming from a source. The contribution of the source is drawn from a Poisson distribution and therefore, it allows the estimation of its location and subsequently of its strength as presented in [47], while it has been effective against fluctuating background as discussed in [48].

6.2.5 Anomaly Detection

Monitoring of radiation levels implies the recording of radiation data and the identification of anomalies in them. By anomalies, we refer to measurements that are substantially higher or lower than the usually observed radiation level. For instance, a sudden elevation in the number of registered radiation counts designates the possibility of the presence (or move) of a new source contributing to the measurement. This kind of abrupt change in count rate is characterized as anomalies and the detection of their onset is of high interest to nuclear security. Notably, anomaly detection in radiation data is mainly tackled as trend or abrupt change identification problems. To that end, several ML techniques have been proposed and tested for anomaly detection in radiation data—either in spectra or time series—and are discussed below.

Bayesian Approaches: A plethora of methods in anomaly detection in radiation data encompass the use of Bayesian inference. The strength of Bayesian approaches lies in the fact that via priors (see Eq. (6.2)) previous knowledge is incorporated into the model. Furthermore, Bayesian approaches allow for the incorporation of recently observed data, thus, making them an attractive choice for real-time anomaly identification either in spectral or time-series data.

Spectral anomalies using a *Bayes Model Selection* method that works synergistically with models of radiation measurements are presented in [49]. Another method that uses the *Bayesian framework of Gaussian processes* for background prediction and anomaly detection is presented in [50]. Similarly, an ML system that utilizes *Bayesian driven Gaussian processes* for anticipating and modeling the background radiation and subsequently for identifying anomalies in time series is discussed in [51]. Furthermore, a *Naïve Bayes classifier* for identifying benign and anomalous data in gamma-ray spectra was proposed in [52].

Principal Component Analysis: Another method for identifying anomalies in radiation spectral data utilizes principal component analysis (PCA). PCA is a nonparametric method that uses a dataset of observations comprised of correlated variables into a set of uncorrelated variables. The uncorrelated variables may be used for the extraction of useful information such as the radionuclide signatures embedded in the data.

PCA has been applied in spectral data taken with a low-resolution radiation detector (more particularly, with a NaI(Tl) gamma-ray detector [5]) and has been proved to be efficient as presented in [53].

Artificial Neural Networks: Several forms of ANN have been proposed for anomaly detection in radiation data with the vast majority of them focusing on the analysis of gamma-ray measurements. The general framework of using ANN in anomaly detection is very similar to the one depicted in Figs. 6.4 and 6.5 and is presented in Fig. 6.10.

In [54], a linear auto-associative memory neural network has been proposed for identifying anomalies associated with a small prespecified number of radionuclides, while in [55] a multilayer perceptron ANN was presented for obtaining the trend of

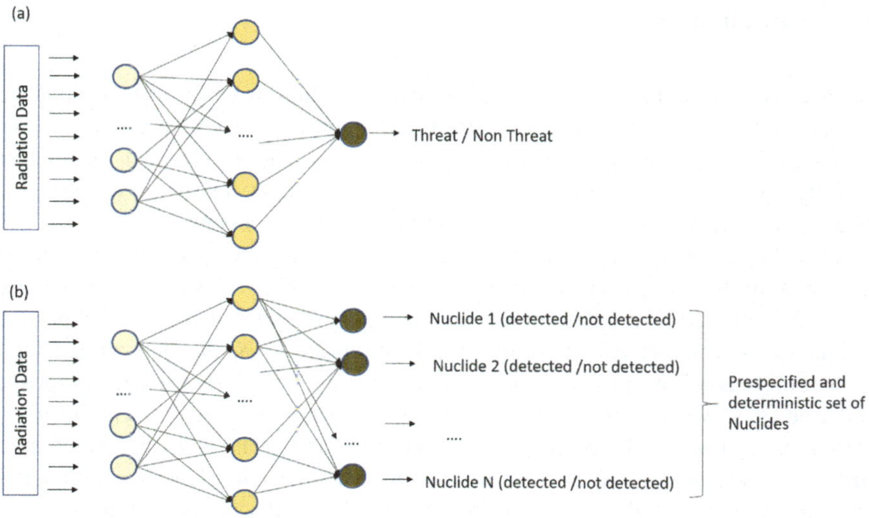

Fig. 6.10 Visualization of ANN for anomaly detection by **a** characterizing a dataset as a threat or no threat, **b** indicating the detection or not of a prespecified set of nuclides

the spectrum and identify anomalies in it. Furthermore, multilayer perceptrons were also used in [56] for identifying anomalies based on the ratio of uranium concentration in a spectrum.

Dynamic Quantum Clustering (DQC): Unsupervised learning has also been applied for anomaly detection in unstructured radiation data. In specifically, a novel approach employing dynamic quantum clustering has been applied to anomaly detection in gamma-ray spectra applicable to radioactive source searches. In brief, DQC is a clustering method that associates every data point in a Euclidean space of n dimensions with a Gaussian wave function:

$$f_i(x) = e^{-(x-x_i)/2\sigma^2} \tag{6.7}$$

where x_i denotes the value of the datapoint I and σ^2 is the variance of the training data. In the next step, the functions in (7) are aggregated over all datapoints constructing a distribution function. DQC considers Eq. (6.7) as a probability amplitude as is the case in quantum mechanics, and therefore, it allows each datapoint to be associated with a cluster center [57].

To that end, the work in [58] employed DQC for anomaly detection in gamma-ray spectral data. In particular, the DQC approach that has been taken is to cluster the data to detect or not the presence of two nuclides, and more specifically Cs-137 and Co-60, i.e., in other words, DQC was looking for specific anomalous data related to those two nuclides.

6.3 Conclusion

Nuclear security is a highly critical area demanding solutions for its problems that are fast, accurate, and adaptive. The field of data analytics merged with machine learning comes to provide intelligent data analytics solutions that are efficient for nuclear security applications. Admittedly, there is a high demand for more sophisticated data analysis methods that are universal and less problem-dependent.

This chapter presented a survey of the areas of nuclear security where machine learning tools have been developed and discussed the problems associated with these areas. It identified four main areas, namely, signature identification, background estimation, sensor placement, source localization, and anomaly detection. Under each of these areas, we provided the machine learning tools that have been proposed and tested to solve the various problems. It should be noted that there is a high variety of tools being developed to propose solutions in nuclear security areas, but this chapter strictly focused on tools that are identified to implement machine learning approaches given the high potential of ML tools in providing fast and accurate solutions.

We aspire that this chapter will provide a decent grasp of the ML tools in nuclear security and aspires to serve as an introduction to the area for newcomers. The development and advancements in developing more sophisticated ML tools will be built only upon the understanding and extending of the current tools.

Acknowledgements This chapter was developed under the auspices of the *Consortium on Nuclear Security Technologies* (CONNECT) supported by the Department of Energy/National Nuclear Security Administration under Award Number(s) DE-NA0003948. Argonne National Laboratory's work was supported by the U.S. Department of Energy, National Nuclear Security Administration, under contract DE-AC02-06CH11357.

References

1. M. Zenko, Intelligence estimates of nuclear terrorism. Ann. Am. Acad. Pol. Soc. Sci. **607**(1), 87–102 (2006)
2. B. Zellen, Rethinking the unthinkable: nuclear weapons and the war on terror. Strat. Insights**1** (2004)
3. J.A. Boscarino, C.R. Figley, R.E. Adams, Fear of terrorism in New York after the September 11 terrorist attacks: implications for emergency mental health and preparedness. Int. J. Emerg. Ment. Health **5**(4), 199 (2003)
4. P.R. Miles, J.A. Cook, Z.V. Angers, C.J. Swenson, B.C. Kiedrowski, J. Mattingly, R.C Smith, Radiation source localization using surrogate models constructed from 3-D Monte Carlo transport physics simulations. Nucl. Technol. 1–17 (2020)
5. N. Tsoulfanidis, S. Landsberger, *Measurement and Detection of Radiation* (CRC Press, Boca Raton, FL, 2010)
6. M.H. Jeong, C.J. Sullivan, S. Wang, Complex radiation sensor network analysis with big data analytics, in *2015 IEEE Nuclear Science Symposium and Medical Imaging Conference (NSS/MIC)* (IEEE, 2015), pp. 1–4
7. D.S. Hochbaum, B. Fishbain, Nuclear threat detection with mobile distributed sensor networks. Ann. Oper. Res. **187**(1), 45–63 (2011)

8. M. Berthold, D.J. Hand, *Intelligent Data Analysis*, vol. 2 (Springer, Berlin, 2003)
9. M. Alamaniotis, J. Young, L.H. Tsoukalas, Assessment of fuzzy logic radioisotopic pattern identifier on gamma-ray signals with application to security, in *Research Methods: Concepts, Methodologies, Tools, and Applications* (IGI Global, 2015), pp. 1052–1071
10. Z. Varga, J. Krajkó, M. Peńkin, M. Novák, Z. Eke, M. Wallenius, K. Mayer, Identification of uranium signatures relevant for nuclear safeguards and forensics. J. Radioanal. Nucl. Chem. **312**(3), 639–654 (2017)
11. M.D. Assunção, R.N. Calheiros, S. Bianchi, M.A. Netto, R. Buyya, Big data computing and clouds: trends and future directions. J. Parallel Distrib. Comput. **79**, 3–15 (2015)
12. L. Holbrook, M. Alamaniotis, A good defense is a strong DNN: defending the IoT with deep neural networks, in *Machine Learning Paradigms* (Springer, Cham, 2020), pp. 125–145
13. T. Burr, M. Hamada, Radio-isotope identification algorithms for NaI γ spectra. Algorithms **2**(1), 339–360 (2009)
14. M. Alamaniotis, T. Jevremovic, Hybrid fuzzy-genetic approach integrating peak identification and spectrum fitting for complex gamma-ray spectra analysis. IEEE Trans. Nucl. Sci. **62**(3), 1262–1277 (2015)
15. J. Mattingly, D.J. Mitchell, A framework for the solution of inverse radiation transport problems. IEEE Trans. Nucl. Sci. **57**(6), 3734–3743 (2010)
16. C.M. Bishop, *Pattern Recognition and Machine Learning* (Springer, 2006)
17. M. Kamuda, J. Stinnett, C.J. Sullivan, Automated isotope identification algorithm using artificial neural networks. IEEE Trans. Nucl. Sci. **64**(7), 1858–1864 (2017)
18. S. Jhung, S. Hur, G. Cho, I. Kwon, A neural network approach for identification of gamma-ray spectrum obtained from silicon photomultipliers. Nucl. Instrum. Methods Phys. Res. Sect. A: Accel., Spectrometers, Detect. Assoc. Equip. **954**, 161704 (2020)
19. J. Kim, K. Park, G. Cho, Multi-radioisotope identification algorithm using an artificial neural network for plastic gamma spectra. Appl. Radiat. Isot. **147**, 83–90 (2019)
20. E. Yoshida, K. Shizuma, S. Endo, T. Oka, Application of neural networks for the analysis of gamma-ray spectra measured with a Ge spectrometer. Nucl. Instrum. Methods Phys. Res., Sect. A **484**(1–3), 557–563 (2002)
21. L. Chen, Y.X. Wei, Nuclide identification algorithm based on K-L transform and neural networks. Nucl. Instrum. Methods Phys. Res., Sect. A **598**(2), 450–453 (2009)
22. C. Bobin, O. Bichler, V. Lourenço, C. Thiam, M. Thévenin, Real-time radionuclide identification in γ-emitter mixtures based on spiking neural network. Appl. Radiat. Isot. **109**, 405–409 (2016)
23. G. Daniel, F. Ceraudo, O. Limousin, D. Maier, A. Meuris, Automatic and real-time identification of radionuclides in gamma-ray spectra: a new method based on convolutional neural network trained with synthetic data set. IEEE Trans. Nucl. Sci. **67**(4), 644–653 (2020)
24. F. Li, J. Wang, L. Ge, F. Hu, F. Cheng, K. Sun, Research on gamma spectrum semi-quantitative analysis based on convolutional neural network. J. Phys.: Conf. Ser. **1423**(1), 012005 (2019). (IOP Publishing)
25. M. Alamaniotis, A. Heifetz, A.C. Raptis, L.H. Tsoukalas, Fuzzy-logic radioisotope identifier for gamma spectroscopy in source search. IEEE Trans. Nucl. Sci. **60**(4), 3014–3024 (2013)
26. M.A. Hogan, S. Yamamoto, D.F. Covell, Multiple linear regression analysis of scintillation gamma-ray spectra: Automatic candidate selection. Nucl. Inst. Methods **80**(1), 61–68 (1970)
27. D.F. Covell, M. Brown, S. Yamamoto, Multiple linear regression analysis scintillation gamma-ray spectra: Theoretical and practical considerations. Nucl. Inst. Methods **80**(1), 55–60 (1970)
28. M. Alamaniotis, C.K. Choi, L.H. Tsoukalas, A new approach in gamma ray spectra analysis: automated integration of peak detection and spectrum fitting using fuzzy logic and multiple linear regression. Am. Nucl. Soc. Meet. Trans. **112**(1), 260–263 (2015)
29. W.R. Russ, Library correlation nuclide identification algorithm. Nucl. Instrum. Methods Phys. Res., Sect. A **579**(1), 288–291 (2007)
30. R. Estep, C. McCluskey, B. Sapp, The multiple isotope material basis set (MIMBS) method for isotope identification with low-and medium-resolution gamma-ray detectors. J. Radioanal. Nucl. Chem. **276**(3), 737–741 (2008)

31. M. Alamaniotis, A. Heifetz, A. Raptis, L.H. Tsoukalas, Fuzzy logic radio isotope identifier for gamma spectra analysis in source search applications, in *American Nuclear Society Annual Meeting* (Chicago, IL, USA, 2012), pp. 211–212
32. Z. Wu, B. Wang, J. Sun, Design of radionuclides identification algorithm based on sequence bayesian method, in *IOP Conference Series: Materials Science and Engineering*, vol. 569, no. 5 (IOP Publishing, 2019), p. 052047
33. Y. Altmann, A. Di. Fulvio, M.G. Paff, S.D. Clarke, M.E. Davies, S. McLaughlin, S.A. Pozzi, Expectation-propagation for weak radionuclide identification at radiation portal monitors. Sci. Rep. **10**(1), 1–12 (2020)
34. J. Kim, K.T. Lim, K. Ko, E. Ko, G. Cho, Radioisotope identification and nonintrusive depth estimation of localized low-level radioactive contaminants using Bayesian inference. Sensors **20**(1), 95 (2020)
35. Z. Wu, B. Wang, J. Sun, Design of radionuclides identification algorithm based on sequence Bayesian method, in *2nd International Conference on Advanced Materials, Intelligent Manufacturing and Automation—Machine Learning and Algorithms*, vol. 569, no. 5 (IOP Publishing, 2019), p. 052047
36. M.W. Swinney, D.E. Peplow, B.W. Patton, A.D. Nicholson, D.E. Archer, M.J. Willis, A methodology for determining the concentration of naturally occurring radioactive materials in an urban environment. Nucl. Technol. **203**(3), 325–335 (2018)
37. C.E. Rasmussen, C. Williams, *Gaussian Processes for Machine Learning* (MIT Press, Boston, 2006)
38. M. Alamaniotis, J. Mattingly, L.H. Tsoukalas, Kernel-based machine learning for background estimation of NaI low-count gamma-ray spectra. IEEE Trans. Nucl. Sci. **60**(3), 2209–2221 (2013)
39. M. Alamaniotis, C. Choi, L.H. Tsoukalas, Data driven modeling of radiation background using an ensemble of learning methods: initial concepts and preliminary results, in *Transactions of the American Nuclear Society Annual Meeting* (2015), pp. 249–252
40. M. Alamaniotis, A data driven methodology for estimation of background spectrum utilizing paired machine learning tools. Transactions **121**(1), 578–581 (2019)
41. M. Alamaniotis, C.K. Choi, L.H. Tsoukalas, Short-term gamma background anticipation using learning Gaussian processes, in *2015 IEEE Nuclear Science Symposium and Medical Imaging Conference (NSS/MIC)* (IEEE, 2015), pp. 1–4
42. M. Alamaniotis, Predicting background count rate of a mobile detector using an optimal ensemble of learning kernel machines, in *American Nuclear Society Annual Meeting, Virtual Conference* (2020), pp. 185–188, 7–11 June 2020
43. J.A. Kulisek et al., Real-time airborne gamma-ray background estimation using NASVD with MLE and radiation transport for calibration. Nucl. Instrum. Methods Phys. Res., Sect. A **784**, 287–292 (2015)
44. J.M. Kirkpatrick, B.M. Young, Poisson statistical methods for the analysis of low-count gamma spectra. IEEE Trans. Nucl. Sci. **56**(3), 1278–1282 (2009)
45. K. Schmidt, R.C. Smith, J. Hite, J. Mattingly, Y. Azmy, D. Rajan, R. Goldhahn, Sequential optimal positioning of mobile sensors using mutual information. Stat. Anal. Data Min.: ASA Data Sci. J. **12**(6), 465–478 (2019)
46. J.M. Hite, J.K. Mattingly, K.L. Schmidt, R. Ştefănescu, R. Smith, Bayesian metropolis methods applied to sensor networks for radiation source localization, in *2016 IEEE International Conference on Multisensor Fusion and Integration for Intelligent Systems (MFI)* (IEEE, 2016), pp. 389–393
47. E.W. Bai, K. Yosief, S. Dasgupta, R. Mudumbai, The maximum likelihood estimate for radiation source localization: initializing an iterative search, in *53rd IEEE Conference on Decision and Control* (IEEE, 2014), pp. 277–282
48. E. Bai, A. Heifetz, P. Raptis, S. Dasgupta, R. Mudumbai, Maximum likelihood localization of radioactive sources against a highly fluctuating background. IEEE Trans. Nucl. Sci. **62**(6), 3274–3282 (2015)

49. D.M. Pfund, Radiation anomaly detection and classification with Bayes model selection. Nucl. Instrum. Methods Phys. Res., Sect. A **904**, 188–194 (2018)
50. M. Alamaniotis, C.K. Choi, L.H. Tsoukalas, Anomaly detection in radiation signals using kernel machine intelligence, in *2015 6th International Conference on Information, Intelligence, Systems and Applications (IISA)* (IEEE, 2015), pp. 1–6
51. M. Alamaniotis, A. Heifetz, A machine learning approach for background radiation modeling and anomaly detection in radiation time series pertained to nuclear security, in *Winter Meeting and Technology Expo* (Chicago, IL, USA, 2020), pp. 477–480, 15–19 Nov 15–19 2020
52. S. Sharma, C. Bellinger, N. Japkowicz, R. Berg, K. Ungar, Anomaly detection in gamma ray spectra: a machine learning perspective, in *2012 IEEE Symposium on Computational Intelligence for Security and Defence Applications* (IEEE, 2012), pp. 1–8
53. R.C. Runkle, M.F. Tardiff, K.K. Anderson, D.K. Carlson, L.E. Smith, Analysis of spectroscopic radiation portal monitor data using principal components analysis. IEEE Trans. Nucl. Sci. **53**(3), 1418–1423 (2006)
54. P. Olmos, J.C. Diaz, J.M. Perez, G. Garcia-Belmonte, P. Gomez, V. Rodellar, Application of neural network techniques in gamma spectroscopy. Nucl. Instrum. Methods Phys. Res., Sect. A **312**(1–2), 167–173 (1992)
55. L.J. Kangas, P.E. Keller, E.R. Siciliano, R.T. Kouzes, J.H. Ely, The use of artificial neural networks in PVT-based radiation portal monitors. Nucl. Instrum. Methods Phys. Res., Sect. A **587**(2–3), 398–412 (2008)
56. V. Vigneron, J. Morel, M.C. Lepy, J.M. Martinez, Statistical modelling of neural networks in γ-spectrometry. Nucl. Instrum. Methods Phys. Res., Sect. A **369**(2–3), 642–647 (1996)
57. M. Weinstein, D. Horn, Dynamic quantum clustering: a method for visual exploration of structures in data. Phys. Rev. E **80**(6), 066117 (2009)
58. M. Weinstein, A. Heifetz, R. Klann, Detection of nuclear sources in search survey using dynamic quantum clustering of gamma-ray spectral data. Eur. Phys. J. Plus **129**(11), 239 (2014)

Chapter 7
AI for Cybersecurity: ML-Based Techniques for Intrusion Detection Systems

Dilara Gumusbas and Tulay Yildirim

Abstract Through this chapter, problems in cybersecurity and the potential AI-based solutions are first introduced. Then, several proposed methods based on Machine Learning in cybersecurity are discussed with examples to give detailed insights to the reader. Finally, the chapter is concluded with an overview of open topics as well as potential directions in cybersecurity within the scope of conclusions of researches in the literature.

7.1 Introduction

With the increasing pace of developments in the digital age, accessing and transferring great amounts of data via an internet connection and evolving cyber threats, cybersecurity-related issues have been increased. Therefore, these issues created the need to deploy more trustable cybersecurity systems, which are composed of a variety of preventative methods.

As one of the widely-studied branches of cybersecurity systems, Intrusion Detection System (IDS) is developed to detect cyber threats and to ensure safe user access and privacy protection. IDS primarily gathers data and makes a detection system into work to catch and identify possible threats for the use of security analysts. Besides, it can be categorized into two: Network Intrusion Detection System (NIDS) and Host-based Intrusion Detection System (HIDS). While NIDS is based on network traffic data that consists of whole interaction among devices on a network, HIDS is based on HIDS agent data collected from only host devices such as operating system logs.

A variety of algorithms are used for IDS which can be observed under three categories: rule-based, statistics-based and Machine Learning (ML) based algorithms. While rule-based algorithms use data distributions to construct a rule and execute it, statistics-based algorithms benefit from previous attack patterns to estimate a sta-

D. Gumusbas (✉) · T. Yildirim
Electronics and Communication Engineering, Yildiz Technical University, Istanbul, Turkey
e-mail: dilara.gumusbas@btu.edu.tr

G. A. Tsihrintzis et al. (eds.), *Advances in Machine Learning/Deep Learning-based Technologies*, Learning and Analytics in Intelligent Systems 23,
https://doi.org/10.1007/978-3-030-76794-5_7

tistical distribution and employ this distribution to detect attacks. The last category falls under ML-based algorithms as a sub-field of Artificial Intelligence (AI), which refers to machines that mimic human cognitive abilities which vary from perception to problem-solving. ML-based algorithms concentrate on the learning part of these cognitive abilities. After learning, they perform classifier training to detect anomalies including known attacks. Since each algorithm has its advantages and disadvantages thereby, choosing the best one depends on the problem and the trade-offs. For example, rule-based algorithms can be performed simply and fast. However, it cannot perform well under missing and/or imprecise data. Moreover, updating this approach is burdensome. Similarly, statistics-based algorithms solve these problems but as a trade-off, they demand high computational power and they are not suitable for large amounts of data. Unlike rule-based and statistics-based algorithms, ML-based algorithms are proposed to solve these problems using inference models that can capture the complexity and can be trained on big data.

7.1.1 Why Does AI Pose Great Importance for Cybersecurity?

As many organizations have started to employ interconnected systems, the amount of data collected and transferred over a network has been growing gradually. Therefore, the protection of the data coming from these systems has become even more fragile to not only unauthorized access but also authorized access by the insider attackers. Moreover, there may be a lack of human force to protect these systems in real-time. To solve these issues, Machine Learning (ML) based approaches, in particular Deep Learning (DL) methods, are frequently used in cybersecurity for three main reasons. First, these approaches are successful to find underlying patterns of data not only for known but also for novel attacks to automate threat and anomaly-based security monitoring and detection. Second, ML-based approaches are good at reducing false positives and lessen the number of false alarms to be analyzed by security analysts. Therefore, they facilitate the process for security analysts and increase their productivity regarding intrusion detection and response time. As a result, reducing the amount of data to be investigated makes a huge contribution to real-time detection and averts data losses. Third, they make the monitoring and detection systems computationally inexpensive and adaptable to update towards evolving attack types. Besides, ML-based approaches are able to predict anomalies.

7.1.2 Contribution

Throughout this chapter, the most popular and up-to-date ML-based solutions for cybersecurity, specifically the ones that focus on DL methods, are discussed. This chapter not only presents an all-inclusive overview of Machine Learning (ML) approaches for cybersecurity by analyzing concerning evaluation results and limitations but also a further investigation on factors that affect the reliability and scalability of these approaches are provided for potential future directions.

7.2 ML-Based Models for Cybersecurity

In this chapter, a variety of selected ML-based approaches for cybersecurity are discussed with their pros and cons. This selection is done with respect to their citation statistics and their contribution as a pioneer to the literature.

Taxonomy for ML-based approaches is also can be found in Fig. 7.1. As can be seen, only Adaboost and Random Forest (RF) algorithms are chosen among a variety of different ensemble methods to discuss for two reasons. First, RF is one of the top algorithms that model intrusions and even compete with the novel deep ML-based models. Second, Adaboost is one of the frequently-used and effective methods for cybersecurity between boosting and bagging methods.

7.2.1 K-Means

K-means is a method where each input data is grouped into randomly chosen clusters (k) according to their distance to these cluster centers and cluster centers are updated until certain criterion such as minimizing distances among clusters is met [1]. Although this algorithm is easy to implement, fast and computationally inexpensive for big data, there are several issues related to the k-means algorithm. First, choosing the optimal number of clusters is difficult. Second, noise in input data has a strong effect on performance results. The last, k-means is negatively affected when different classes create the same cluster due to their same mean or data is non-convex.

Several methods have been proposed by using different distance metrics. For instance, the method introduced in [2] uses System Call Frequency Distribution (SCFD) to calculate similarity metrics with k-means, where the cut-off distance of clusters is calculated using cumulative distribution function (CDF) with Mahalanobis distance to differentiate normal from attacks. The method achieves better detection than Euclidean distance on outliers on the private dataset used while it may not detect local variations in system call sequences. Similarly, the method introduced in [3] employs k-means with Gaussian Similarity Measure on DARPA98 dataset.

To achieve high detection rates, several approaches employed hybrid methods. For example, the model proposed in [4] first employs k-means to obtain different training subsets then uses five Fuzzy Neural Networks (NN). As a final step of the model, it classifies with SVM. The model achieves high detection results on KDD99 for each attack type. Similar to [4], the model proposed in [5] selects the most distinctive data samples with k-means then classifies these samples with NN. However, the model achieves low detection rates for minority classes: R2L and U2R in KDD99. The method introduced in [6] uses k-means as a first step of separating data into clusters

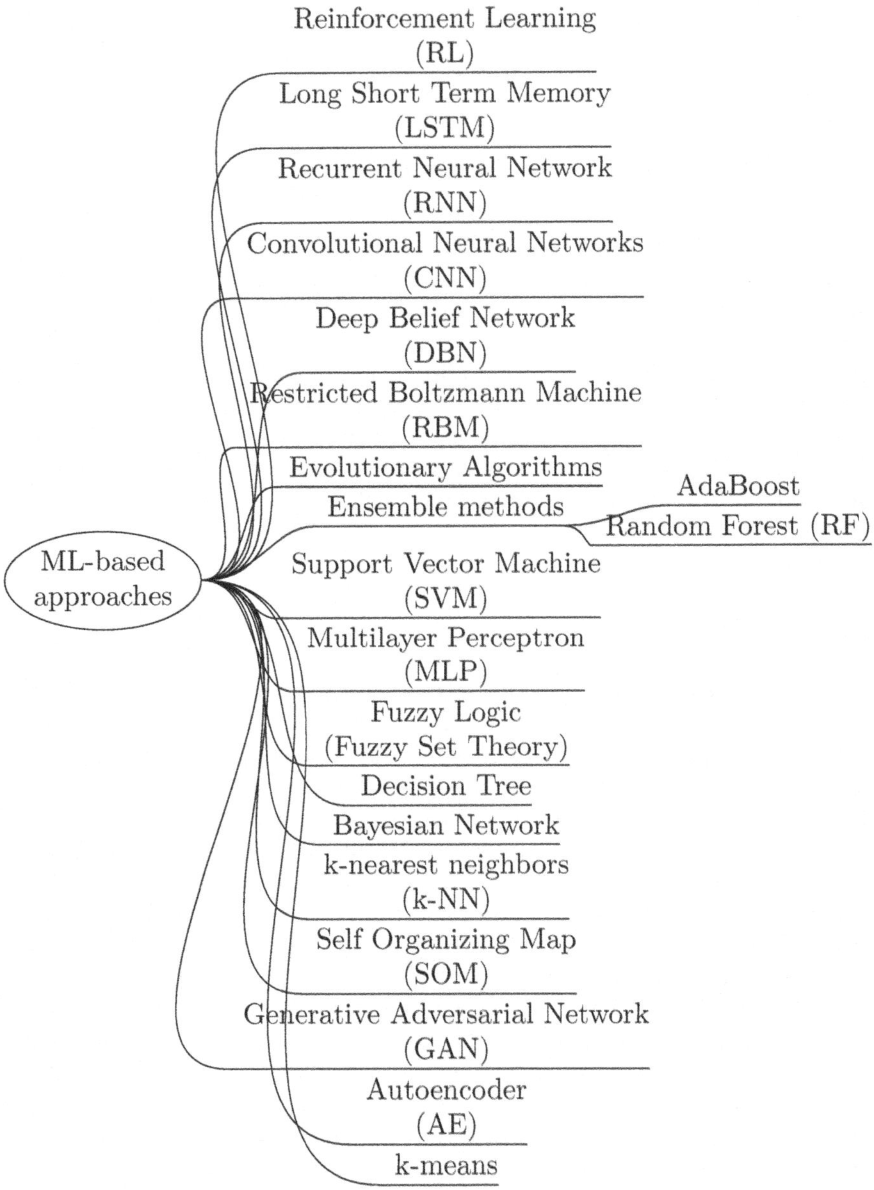

Fig. 7.1 Taxonomy of ML-based algorithms for cybersecurity

then learns subgroups in clusters with C4.5 decision tree while the method proposed in [7] does the same with Naive Bayes. Both methods achieve a high true positive rate with low false positives on the KDD99 dataset.

7.2.2 *Autoencoder (AE)*

AE is a type of unsupervised DL method that first encodes then decodes the original data to build novel representations of the data, which can be seen in Fig. 7.2. While the encoding layers make data representations into lower dimensions to find the most informative feature space, decoding layers samples from this space to original feature space under an unsupervised fashion [8]. All weights are optimized by minimizing reconstruction error. AE has generally been used dimension reduction for cybersecurity thanks to its capability of extracting informative feature representations. However, choosing the optimal structure of encoding and decoding layers is difficult.

Several works are published for two different combinations of AE: AE with shallow/deep ML algorithms and AE with statistical algorithms or statistics-driven AE models such as VAE with shallow ML algorithms. The studies proposed for the first combination in [9, 10] use AE for dimension reduction/nonlinear feature extraction and combine it with several shallow classifiers such as SVM on the NSL-KDD dataset. It is reported that combinations with AE achieve higher accuracy compared

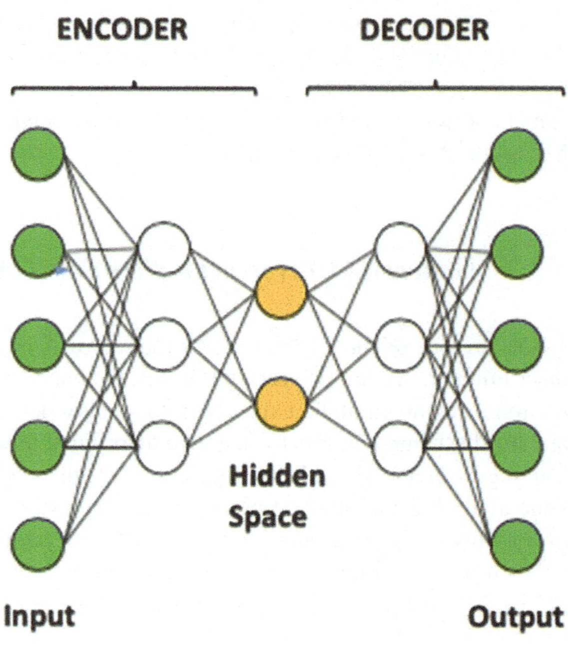

Fig. 7.2 An example of an autoencoder

to the combinations with other dimension reduction methods. Furthermore, the study presented in [11] employs AE with a softmax regression classifier on the same dataset and reports higher accuracy than the previous ones. In addition to 1-hidden-layered AE, multi-layered versions, also known as stacked AEs, are used in the works introduced in [12–14]. These works combine stacked AEs with shallow classifiers. While the first two uses random forest on NSL-KDD and KDD99 datasets, the latter uses a radial basis function to achieve high overall accuracy on the AWID2018 dataset. Similarly, the model introduced in [15] uses stacked AEs to extract valuable information from raw traffic data and automate the intrusion detection process. Besides, the works proposed in [16, 17] first use AE to extract meaningful information from raw network traffic then detects anomalies with one of the deep ML algorithms, CNN.

The methods proposed for the second combination of AE in [18, 19] use VAE to reduce dimension for raw network traffic and several featured datasets named NSL-KDD and UNSW-NB15 datasets, respectively. In the second work, several shallow algorithms such as random forest are also used to detect anomalies using the output of VAE. In a similar manner, the work introduced in [20] employs VAE with gradient-based linear SVM to detect some particular attacks on the AWID2019 dataset, where SVM first reduces feature dimension then VAE selects the most relevant features. It is reported that the detection rate is higher than state-of-art models. In addition to models in [18–20], the model introduced in [21] combines VAE with GAN and DNN. Basically, it uses VAE to obtain new input representations formed in a statistical and nonlinear way then GAN to make less-represented intrusions augmented. Finally, the model uses DNN to classify unknown ones as well as known intrusions. Furthermore, several AE combinations with statistical algorithms in [22, 23] adopt AE to extract nonlinear representations, then use density estimation on NSL-KDD and Gaussian Mixture Model (GMM) on KDD99, respectively. The results are indicated that AE with statistical algorithms improves detecting intrusions that can be highly detected by frequency related features. Similarly, the work presented in [24] combines AE with statistical models and achieves higher accuracy than state-of-art deep and shallow ML on the NSL-KDD dataset.

7.2.3 Generative Adversarial Network (GAN)

GANs are one of the DL algorithms that consists of an encoder, a generator and a discriminator. As can be seen in Fig. 7.3, the encoder first extracts statistical information from the input, then the generator creates new samples using the information and the discriminator tries to differentiate original input from created ones [25]. The training of generator and discriminator is frequently done to minimize loss of the generator while maximizing loss of the discriminator. GANs have a great advantage of not only classifying but also augmenting new data samples only using statistical characteristics of input, in particular for minority classes in a dataset. Therefore, it has gained great interest in cybersecurity applications.

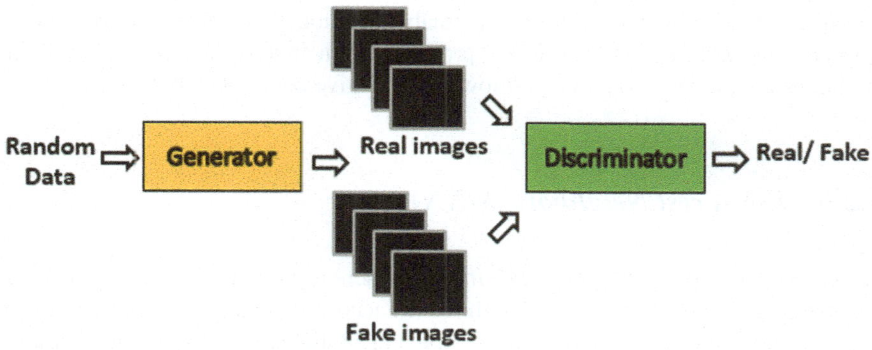

Fig. 7.3 An example of a generative adversarial network

Several studies are done on data augmentation for cybersecurity datasets which have frequently imbalanced data samples among classes. The work proposed in [26] uses sequence GANs to augment the ADFA-LD dataset. Similarly, the studies introduced in [21, 27] employ GANs to augment raw network traffic data. Both report an improvement in detection results. Similarly, another method proposed in [28] uses Flow Wasserstein GANs to generate adversarial data samples then employs them to detect and model anomalies better for cybersecurity. The evaluation is conducted on ISCX-2012 and ISCX-2017 datasets and an outperforming detection rate is reported.

In addition to the use of GANs for data augmentation, GANs are used for the classification of anomalies. The study proposed in [29] uses GANs for anomaly detection. Similarly, the work proposed in [30] employs GANs with some modifications to achieve an improvement in time. Both reported high and similar detection rates on the KDD99 dataset.

7.2.4 Self Organizing Map

Self Organizing Map (SOM) is an ML method where input data is organized and reduced into a lower dimension in an unsupervised manner with a competitive learning algorithm [31]. Even though SOM is one of the easiest methods to use, it is unstable to distribution shifts in the input data as well as the initialization of neuron weights.

The model introduced in [32] employs hierarchical SOM for a variety of different design structures. The model achieves unsatisfactory detection rates except for DoS attacks on the KDD99 dataset. Several approaches also use hybrid models based on SOM. For example, the model proposed in [33] first reduces feature space with PCA by selecting eight eigenvectors with less noise with Fisher Discriminant Ratio then classifies with SOM. The model achieves high sensitivity and specificity on the NSL-KDD dataset. Another approach proposed in [34] employs J.48 decision tree for misuse detection and a SOM for anomaly detection. This approach first models normal data for TCP, UDP and ICMP protocols, then analyses anomaly with SOM. It obtains a high detection rate with low false positive on the KDD99 dataset.

7.2.5 K-Nearest Neighbors (k-NN)

K-nearest neighbor (k-NN) is an ML method where each input data is assigned to a class of its randomly chosen neighborhood (k) according to their distance similarity [1]. Despite using fewer parameters, simple calculations, scalability, robustness to noise and uncovering natural patterns of data, there are a couple of problems related to this method. First, choosing the right k parameter is not simple. Because too small k value models noise while too big k models other classes. Second, clustering algorithms like k-NN makes the algorithm stuck on a local minimum point. Third, using the Euclidean distance metric might not separate tangled data even it may contribute to misleading results. The last, this method becomes slow and memory inefficient for high dimensional data.

To accelerate detection time as well as to obtain a high detection rate, several approaches combine k-NN with methods such as feature selection, new feature representations. For example, the model introduced in [35] uses Gaussian Mixture Model (GMM) to model statistical regularities in features. After GMM parameters are modeled in Gaussian form, these parameters are fined tuned with the EM algorithm. Then, the model classifies with k-NN using these parameters. The model achieves satisfactory/good detection results in particular on R2L and U2R attack types in the KDD99 dataset. Similarly, the model proposed in [36] first uses a new feature selection approach that assumes the variance of a feature as a quality indicator and reduces all low-quality features. After the selection of ten features, the model uses k-NN and achieves faster detection than the one without feature selection on the KDD99.

Several methods in the literature used k-NN within the cascade system to achieve higher detection rates. For instance, the method proposed in [37] first aggregates of multi-resolution network traffic flow then ranks this aggregated traffic flow according to the level of the anomaly. After ranking, the method classifies a high-level anomaly labeled flow as an intrusion by thresholding. The method achieves sufficient detection accuracy on the KDD99 dataset (more than 90% DR with less than 1% FPR). Similarly, the method described in [38] uses three tiers k-NN based cascade system. The method first extracts cluster centers and nearest neighbors then form training data by summing the calculated distances between data and its cluster center and data and its nearest neighbor. The method obtains a significantly high accuracy and the detection rate on the KDD99 dataset. (99.76% accuracy, 99.99% DR with 0.003 FAR) Similar to methods proposed in [37, 38], the method introduced in [39] uses a two-tier system based on k-NN with the knowledge-based system. The method first uses a knowledge-based system to generate alarms then filters these alarms with k-NN. The method, however, achieves average results on the DARPA-1999 dataset.

7.2.6 *Bayesian Network*

Bayesian Network is an ML-based model that learns from the intrinsic behavior of input data by using statistical dependencies without requiring prior knowledge. Although this network detects small deviations/variations in data and can be applied for continuous, discrete as well as binary input data types, there are some negative aspects related to it. First, it may be fragile towards distributed/low-frequency attacks that create normal-like traffic. Second, it may be ineffective towards correlated features since it assumes that every feature is independent of each other while calculating statistical dependencies. Third, it is slow for larger-scale input data due to calculations.

Bayesian Network is applied for several scenarios introduced in [40–43]. The approach proposed in [40] uses Naive Bayes on NSL KDD and achieves a high true positive rate for DoS, R2L, Probe attacks. Similarly, the model introduced in [43] uses Naive Bayes after discretizing data. However, it only improves DoS attack detection on the KDD99 dataset. Besides, the work introduced in [42] modifies Naive Bayes with Discretize Filter, where a set of predefined intervals are used to change feature values into interval values, and obtains higher detection with a small alarm rate than Naive Bayes itself on NSL KDD.

Several works in the literature combine Bayesian Network with other shallow algorithms to achieve higher detection results. For example, the model proposed in [44] employs Correlation Feature Selection and Information Gain for feature selection then combines Adaptive Boosting and Naive Bayes on NSL KDD dataset to detect anomalies. Similarly, the method introduced in [45] combines Naive Bayes with ADAM based system on DARPA98 and DARPA99 datasets.

7.2.7 *Decision Tree*

Decision Tree is an ML model where all features are scanned and separated into groups. The model is composed of three main elements that are leaf, root and decision node. From roots of decision tree to leaves, if-else command path obtains output while leaving less important features behind [46]. In particular, it is effective for classes with insufficient data and able to work with categorical as well as numerical input data. Moreover, it automates feature selections for trees and can be easily interpretable thanks to the tree structure. However, it ignores the mutual relationships among features. Mostly-known decision tree models are C4.5, CART, J48, respectively.

The model proposed in [47] employs a suffix tree using a sequence covering. This model calculates similarities between system calls on UNM and ADFA-LD datasets. Although it manages free from the length of symbolic sequences with faster-convergence than rival methods such as LCSt, the percentage of normal data samples in the training dataset plays a crucial role. Several models in the literature combine Decision Tree with other algorithms. For example, the model proposed in [48] employs Decision Tree with Information Gain to investigate features and their relevance to each sub-attack type. The method reports that source bytes and destination bytes are two of the most relevant features for all attack types on the KDD99 dataset. Similarly, another hybrid model introduced in [49] uses a Decision Tree with Bayesian clustering. The model first splits data into three classes: DoS, Probe and others then classifies others into the attack and normal. As a final step, the model separates U2R and R2L attacks. The model achieves high detection rates except for U2R and R2L attacks on the KDD99 dataset.

7.2.8 Fuzzy Logic (Fuzzy Set Theory)

Fuzzy logic is a method where classification boundary is treated as soft boundary among the range of 0–1 rather than firm boundary according to fuzzy rules [50]. These rules are defined to classify classes by the experts. Despite having uncertainty flexibility towards input data with fuzzy rules, these rules cannot be scaled to other systems easily.

The model introduced in [51] uses fuzzy association rules based IDS approach. However, the model achieves above average detection rates on the KDD99 dataset. To achieve higher detection rates, several hybrid methods extend their Fuzzy Logic model by combining it with other algorithms. For instance, the models proposed in [52, 53] combine Fuzzy Logic with Genetic Algorithm. Both methods achieve high accuracy and detection rates with low false-positive rates on private, KDD99, NSL KDD and Gure-KddCup datasets. Similarly, several hybrid methods proposed in [54, 55] first employ a fuzzy rough set for feature selection/reduction. Then, [54] uses k-NN and achieves state-of-art detection with a small error rate on KDD99 while [55] creates GMM based attack and normal pattern library and obtains high detection with low error rate on NSL KDD dataset. Another hybrid approach proposed in [56] uses Fuzzy Logic to create different training subsets then employs ANN to classify attacks. The approach achieves improved precision and recalls particularly on R2L and U2R attack types in the KDD99 dataset.

7.2.9 Multilayer Perceptron (MLP)

A Multilayer Perceptron (MLP) is a type of Artificial Neural Network (ANN) that is composed of neurons with associated scalar weights to interconnect other neurons, activation functions and layers. This network uses a backpropagation algorithm such as Gradient Descent to tune/update weights by minimizing classification error [57]. Despite its robustness to noise and being compatible with linear and non-linear inputs, choosing the optimal number of layers and neurons is difficult. Moreover, it may be stacked to local minima resulting from Gradient Descent (Fig. 7.4).

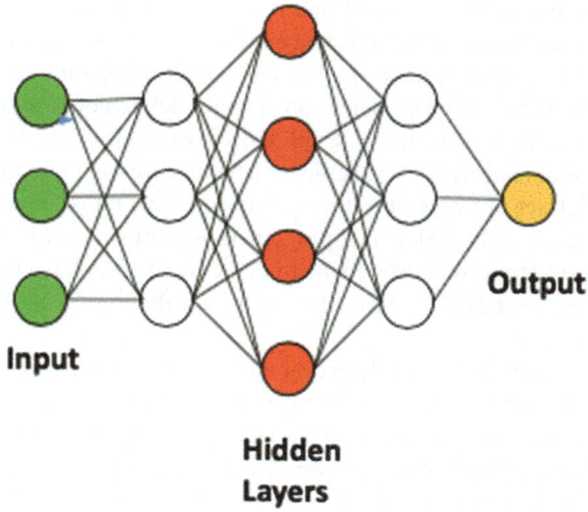

Fig. 7.4 An example of multilayer perceptron [57, 58]

The model introduced in [59] first employs Particle Swarm Optimization (PSO) to optimize parameters of MLP then conducts classification via MLP. The model obtains slightly better error rates than the one without PSO. Another anomaly model proposed in [60] first converts symbolic data to numerical by using Ghosh prototype and the Canberra metric then employs MLP with the chaotic neuron. However, the model obtains average results on the DARPA 1998 dataset.

A misuse based method is proposed in [61]. The method employs three different 3-layer MLP structures and trains each by using ICMP, TCP and UDP based features, separately. Then, the method uses rules by thresholding each output from MLP. Despite obtained high detection rates on the private dataset with known DDoS and unknown DDoS attacks, the model is limited to a few types of attacks and may not handle DoS attacks with encrypted packet headers. Additionally, choosing the right threshold value is not easy.

7.2.10 Support Vector Machine (SVM)

Support Vector Machine (SVM) is an ML method that defines a hyper-plane by maximizing the margin among data samples from different classes [62]. To do that, the method uses the closest data samples to the hyper-plane from different classes and takes advantage of kernel space to map data into higher dimensional space. Despite having the advantage of separating non-linear data using kernel-trick, the choice of kernel type and the volume of feature space due to support vector size have a great impact on performance results.

Several methods proposed in the literature frequently employ SVM with other algorithms or cascade SVM. For instance, the model proposed in [63] uses two different SVMs, where one is for misuse, other is for anomaly detection. Also, the model presented in [64] first employs one of the Manifold methods, k variable locally linear embedding(kv-LLE), and Isomap for feature reduction. Then, the model uses SVM for anomaly detection. The model achieves high detection rates for anomaly detection on KDD CUP 99 and UNM datasets. However, kv-LLE and kv-Isomap combined with SVM achieves a better detection rate than SVM itself on the KDD99 dataset in terms of reducing false-positive rates. In addition to [63, 64], the model proposed in [65] first employs memory-efficient kernel tricked PCA for online feature extraction then uses SVM to classify. The model achieves a high overall detection rate on KDD99. Also, the main contribution of the model is fast real-time/online detection.

Another hybrid detection model with SVM is proposed in [66]. The model uses an agent for anomaly detection and SVM for misuse detection, where four different SVMs are trained for each attack type in the KDD99 dataset. The model achieves a fast and high detection rate. Additionally, the method presented in [67] first reduces the feature space from 41 to 19 using the GRF method then classifies with SVM. The method achieves high accuracy by improving training time.

7.2.11 Ensemble Methods

Ensemble classifiers are a combination of two or more shallow classifiers. Random Forest (RF) is one of the frequently used ensemble classifiers that consists of a bunch of decision trees. Although it is a shallow classifier, training many shallow decision trees contributes to optimizing/generalizing the model, add randomness and prevent from overfitting [68].

Several methods in the literature employ RF. For example, the method introduced in [69] first separates data using known patterns for specific intrusions then decides whether data belong to anomaly by using RF. The model achieves a high overall detection rate with low false alarms/positives on the KDD99 dataset. Similarly, the model proposed in [70] uses an RF-based model named as Hybrid Isolation Forest(HIF). The model first assumes unoccupied areas in feature space as normal then models potential-anomaly-spots using few anomaly samples. The model achieves high AUC rates with a small improvement compared to other rival algorithms such as SVM on ISCX IDS 2012 dataset. Another RF-based model proposed in [71] first preprocesses by using SMOTE (Synthetic Minority Oversampling Technique) to grow the U2R training sample size from 52 to 468 in NSL KDD dataset. Then, the model employs Information Gain to reduce features from 41 to 19. After preprocessing is completed, data is given to RF for multiple classifications. The model achieves state-of-art detection rates without false positives by improving detection for minority attack types. Likewise, the model introduced in [72] first employs RF for misuse detection then uses k-means for anomaly detection. The model obtains high overall detection with low false alarms on the KDD99 dataset.

Besides RF-focused models, there are a variety of shallow classifier combinations with the Adaboost algorithm in the literature. For instance, the model introduced in [73] first uses RF for feature selection then employs k-means++ to separate data into three clusters that represent normal, R2L and U2R attacks and remaining attack types due to similarity among normal, R2L and U2R. After these steps, the model uses Adaboost to separate attack cluster into four sub-attack classes and achieves state-of-art accuracy on balanced KDD99. Similar to [73], the method described in [74] is composed of boosting algorithm, Adaboost and RF. The method achieves identical accuracy to only RF used one. Moreover, the model proposed in [75] employs an ensemble of J48, Naive Bayes, Random Tree, AdaBoost, Meta Papping, Decision-Stump and REPTree on the NSL KDD dataset while another model introduced in [76] uses AdaBoost on the KDD99 dataset. Both models report high accuracy.

In addition to Adaboost combined models discussed, other combinations are done to create an ensemble model. For example, the model proposed in [12] first uses Non-symmetric Deep Autoencoder (NDAE) for dimension reduction then employs RF for classification. It achieves high accuracy on DoS and Probe attacks while obtaining below-average accuracy for minority attacks on both KDD99 and NSL KDD datasets due to the need for more samples to train/tune Deep Learning models well. Moreover, the main contribution of the model is a shorter process time than standard DBN techniques. Similarly, the model proposed in [77] combines CART with Bayesian Network. The model obtains high detection rates especially for DoS, Probe and R2L attacks on the KDD99 dataset. Moreover, a misuse based detection model proposed in [78] employs ensemble boosted decision trees. The model achieves high detection rates only for DoS, R2L and probe attacks on the KDD99 dataset.

7.2.12 Evolutionary Algorithms

Evolutionary classifiers are models that are inspired by the natural process of evolution to solve optimization problems. There are a variety of evolutionary classifiers such as Genetic Algorithms (GA), Ant Colony Optimization, Particle Swarm Optimization (PSO). Among them, GA is one of most used type of evolutionary classifiers that generates chromosomes randomly and make stochastic searches until the best combinations of chromosomes are found. During these searches, chromosomes evolve through mutation, crossover and selection. GA is advantageous at detecting global minima without requiring prior information about feature space. However, the decision of fitness function and hyperparameters is difficult.

The model proposed in [79] first employs PSO to reduce features in the KDD99 dataset to eighteen then classifies with SVM. The model achieves high accuracy with low false positives. Another model proposed in [80] first uses a rule mining process then optimizes with graph-based genetic network programming to model parameters. The model obtains high overall accuracy with low false positives on the KDD99 dataset. Besides, the study introduced in [81] combines two GA with fuzzy sets to evolve new fuzzy rules. It evaluates new rules on several benchmark datasets.

7.2.13 Convolutional Neural Networks (CNN)

CNN is one of ANN types that is composed of different variations of convolutional, pooling and fully connected layers [82]. As can be seen in Fig. 7.5, the input is first processed by convolutional and pooling layers that create a variety of feature maps to find informative representations of the input. Then, it is given to a fully-connected layer to classify. Besides, all weight parameters for convolutional and fully-connected layers are optimized by gradient descent during training.

CNN has a great advantage of automated feature extraction and is frequently used in many recent works. However, using CNN and its powerful backbones in the 2D domain may require an additional step for the preprocessing of 1D input to be compatible with 2D input. For example, several approaches proposed in [83–85] use different preprocessing methods with CNN. The first one converts symbolic features into numeric with binarization while converting continuous features into intervals to make them numeric features. Then, one-hot encoding and 2D reshaping are applied to all converted features to form them pixel-like, respectively. Similar to the first method, the second one takes raw input composed with numbers as 8-bit binary numbers and converts these binary numbers into its analogous decimal pair. Afterward, 2D reshaping is applied to form them image-like. Likewise, the third converts input into grayscale image format after the process for the first method is done. Similarly, several methods employ these preprocessing steps after feature selection is done [86].

In addition to the new preprocessing steps, some works proposed in [87–89] focus on more state-of-art CNN backbones to obtain higher detection rates while others described in [16, 90, 91] give their attention to designing a novel CNN structure for DoS/DDoS detection on KDD99, private and CICDDoS2019 datasets, respectively.

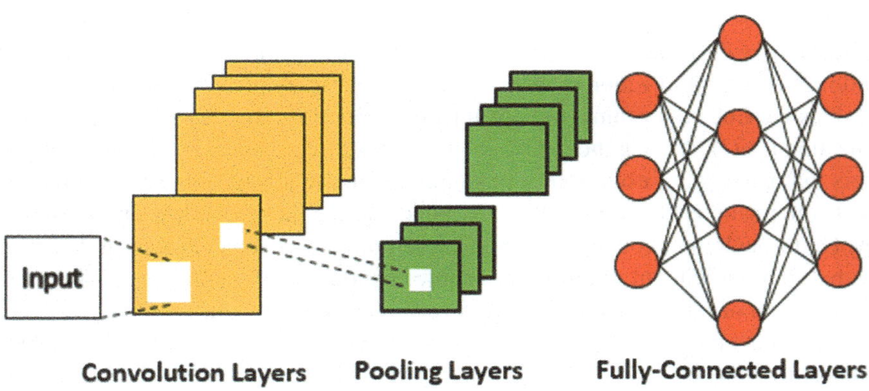

Fig. 7.5 An example of convolutional neural networks

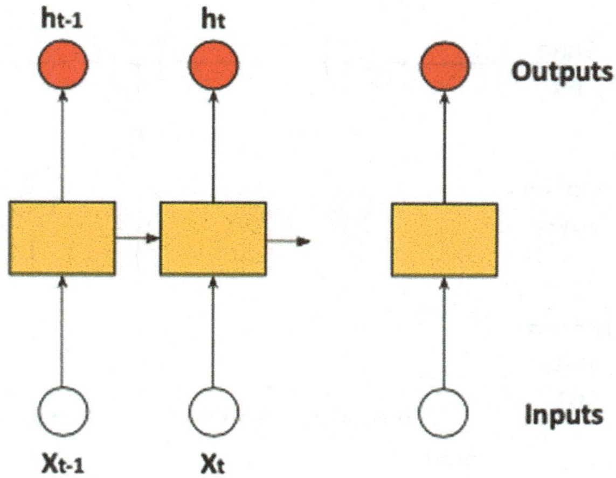

Fig. 7.6 An example of a recurrent neural network [92]

7.2.14 *Recurrent Neural Network (RNN)*

A Recurrent Neural Network (RNN) is a type of ANN in which the hidden neurons are connected by following a temporal sequence. Thanks to such arrangement of their nodes, RNNs are principally used to process data in the form of time series [92]. Even though RNN poses some problems such as vanishing gradients, it is frequently used for time-series modeling for cybersecurity (Fig. 7.6).

The model proposed in [93] uses RNNs and achieves high detection accuracy and fast real-time performance on the DARPA98 dataset. Similarly, the work introduced in [94] obtains higher accuracy than CNN, SVM, and RF classifiers on the ADFA-LD dataset. Another method proposed in [94] employs RNN with Gated Recurrent Unit (GRU) on the ADFA-LD dataset. Since this dataset consists of system calls with various lengths, the semantic model uses different lengths between 10–30 of system-calls. The model achieves high AUC scores. However, finding the optimal length of the system-call sequences may be problematic.

7.2.15 *Long Short Term Memory (LSTM)*

LSTM is designed as an improved version of RNN. An LSTM network consists of sequentially-connected neurons that are composed of input and output gate units, known as a memory cell, to save the memory of previous inputs and forget these inputs for a new interval of time [95]. As can be seen in Fig. 7.7, the input is processed by sequential neurons to model as time-series.

LSTM has a great advantage of modeling time-series and it is employed in many recent works for cybersecurity despite the difficulty of choosing optimal hyperparameters. For instance, the models proposed in [96, 97] employ only LSTM as

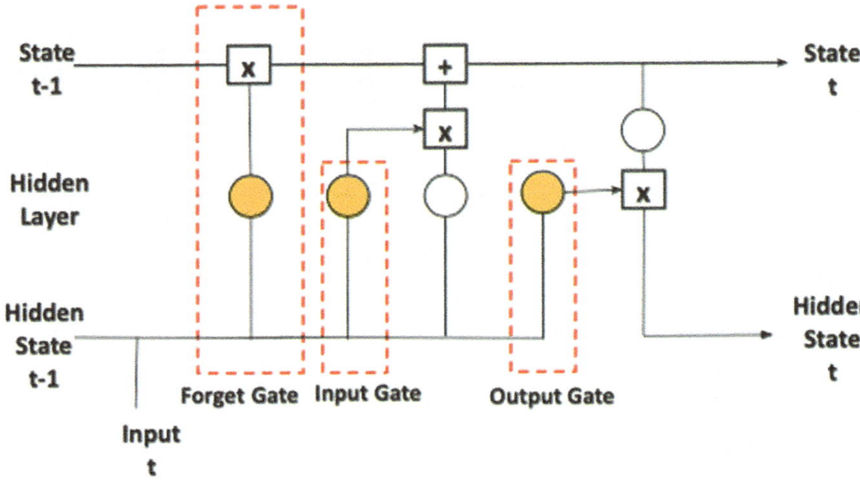

Fig. 7.7 An example of a long short term memory

a classification algorithm on several benchmark datasets under different settings. While the first uses a 3-layer-structure on KDD99, UNM and ADFA-LD datasets, the second cascades LSTM with the voting module and both report high detection accuracy. Similarly, the work introduced in [98] only employs Bidirectional LSTM on the UNSW-NB15 benchmark dataset and the study proposed in [99] adapts multivariate correlations analysis into LSTM on the NSL-KDD dataset to separate feature subsets more efficient.

Besides, several works combine LSTM with DL algorithms, in particular with CNN. The works proposed in [100, 101] combine LSTM with CNN on frequently used benchmark datasets: KDD99 and CICIDS2017, respectively. Similarly, the approach introduced in [102] combines bi-directional LSTM with CNN to extract temporal and spatial features on NSL-KDD and UNSW-NB15 datasets after balancing the datasets with SMOTE.

7.2.16 Restricted Boltzmann Machine (RBM)

RBM is an energy-based neural network with two layers: hidden and visible, where the weights of the network are trained in an unsupervised fashion [103]. Since RBM can extract hidden patterns of input data modeling probability distributions of inputs, it is generally used for feature extraction in cybersecurity (Fig. 7.8).

The study introduced in [105] employs RBM for the FPGA-based intrusion detection system. Using RBM for this system increases computational efficiency by 30% on HTTP CSIC 2010 dataset. Similarly, the works proposed in [106, 107] use RBM for dimension reduction on the KDD99 dataset to improve accuracy and memory

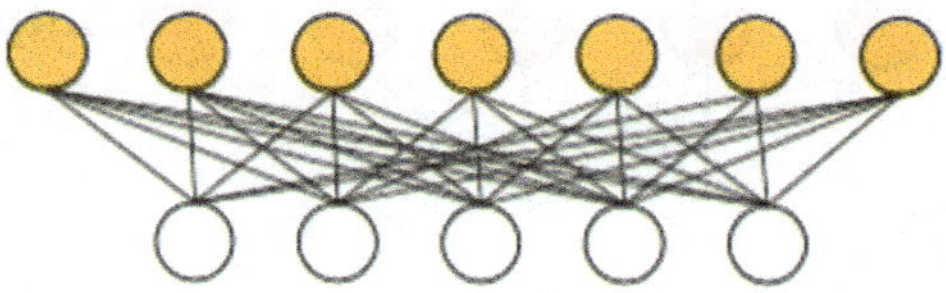

Fig. 7.8 An example of a restricted Boltzmann machine [103, 104]

efficiency. The work introduced in [108] uses RBM with AE on KDD99 to make for feature extraction and dimension reduction process powerful. The evaluation results show an improvement in detection.

7.2.17 Deep Belief Network (DBN)

DBN is composed of several RBM blocks. As can be seen in Fig. 7.9, all layers are fully-connected and only connections between successive layers are allowed while interconnections among layers and within each layer are prohibited [109]. DBN has a great advantage of being time-efficient thanks to training in a greedy-fashion as well as a good feature extractor/selector without requiring any supervision.

DBN is frequently employed for feature selection and combined with other ML algorithms. For example, the method presented in [110] uses DBN as a feature extractor with SVM to classify attack types on the NSL-KDD dataset. In a similar manner, the approaches proposed in [111, 112] employ DBN to model and detect anomalies on the same benchmark dataset. To improve, a novel DBN-based model with extreme learning machine (ELM) on the same dataset is proposed in [113]. This model improves detection while reducing false positives. In addition to the works mentioned, a variety of methods that use DBN are designed on the KDD99 benchmark dataset. For instance; the [114] uses DBN with probabilistic ANN to detect intrusions. Similarly, the models introduced in [115, 116] employ DBN on the same benchmark dataset and report improved detection results compared to shallow ML methods such as ANN and SVM.

7.2.18 Reinforcement Learning (RL)

Reinforcement Learning (RL) is a DL method which uses agent interacting with the environment directly under three concepts: state, action and reward function. The agent first learns from its actions according to the reward function then optimizes its state, where the reward function shows how good the action is and enables the agent to learn what is good action by giving reward [117]. One of the frequently used RL methods is Q-Learning, where the reward function is based on Bellman Equation.

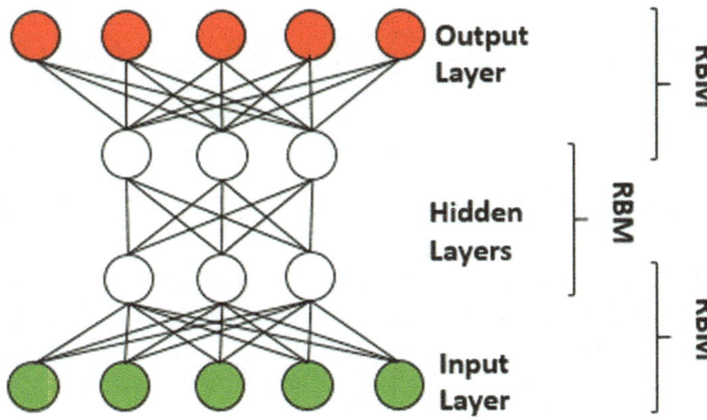

Fig. 7.9 An example of a deep belief network

The model proposed in [118] first uses rough set theory (RST) to reduce features and discretize data with Q-learning that finds optimal cut values for features in the dataset. It achieves high accuracy on the NSL-KDD dataset for anomaly detection.

7.3 Open Topics and Potential Directions

This section presents an overview of open topics and potential directions regarding new feature representations and unsupervised learning-based cybersecurity.

7.3.1 Novel Feature Representations

As new AI-based methods are available, not only security systems take advantage of these methods, but also hackers use them for testing their novel attacks. For instance, GANs are used to generate new samples to train the system better, however, hackers could also take advantage of generating normal traffic data from a variety of sources with different background noise then use their statistical properties to design a new attack.

Besides, these novel attacks can evolve to mimic normal traffic data and cheat the system. For example, DDoS attacks are a low-frequency version of DoS attacks and their characteristics exhibit a great similarity to normal traffic data. Although they are from the same attack family, low-frequency related features are more vital to detect DDoS attacks. Therefore, novel feature representations gain more importance and remain as a potential research area.

7.3.2 *Unsupervised Learning Based Detection Systems*

With the growing usage of internet-based systems, huge amounts of unlabelled data are collected and their labeling process becomes expensive. Moreover, novel attacks may be labeled as normal by a human expert or their novel characteristics may not be differentiated as an intrusion by security systems, that are biased to their training dataset until they cause system breakdowns. Thus, unsupervised AI methods such as recursive Bayesian Network, causal models pose great potential for future researches to extract and detect internal characteristics without any supervision.

References

1. C.M. Bishop, *Pattern Recognition and Machine Learning (Information Science and Statistics)* (Springer, 2006)
2. M.-K. Yoon, S. Mohan, J. Choi, M. Christodorescu, L. Sha, Learning execution contexts from system call distribution for anomaly detection in smart embedded system, in *Proceedings of IoTDI* (2017), pp. 191–196
3. G.R. Kumar, N. Mangathayaru, G. Narsimha, A novel similarity measure for intrusion detection using gaussian function. CoRR abs/1604.07510 (2016)
4. A.M. Chandrasekhar, K. Raghuveer, Intrusion detection technique by using k-means, fuzzy neural network and SVM classifiers, in *2013 International Conference on Computer Communication and Informatics*, Jan 2013, pp. 1–7
5. K. Faraoun, Neural networks learning improvement using the k-means clustering algorithm to detect network intrusions. INFOCOMP J. Comput. Sci. **5**, 28–36 (2006). ISSN: 1807-4545
6. A.P. Muniyandi, R. Rajeswari, R. Rajaram, Network anomaly detection by cascading k-means clustering and c4.5 decision tree algorithm (2012)
7. Z. Muda, W. Mohamed, M.N. Sulaiman, N. Udzir, K-means clustering and Naive Bayes classification for intrusion detection. J. IT in Asia **4**, 13–25 (2016)
8. I. Goodfellow, Y. Bengio, A. Courville, *Deep Learning* (MIT Press, 2016), http://www.deeplearningbook.org
9. B. Abolhasanzadeh, Nonlinear dimensionality reduction for intrusion detection using autoencoder bottleneck features, in *2015 7th Conference on Information and Knowledge Technology (IKT)* (2015), pp. 1–5
10. M. Yousefi-Azar, V. Varadharajan, L. Hamey, U.K. Tupakula, Autoencoder-based feature learning for cyber security applications, in *2017 International Joint Conference on Neural Networks (IJCNN)* (2017), pp. 3854–3861
11. A. Javaid, Q. Niyaz, W. Sun, M. Alam, A deep learning approach for network intrusion detection system, in *Proceedings of the 9th EAI International Conference on Bio-Inspired Information and Communications Technologies (Formerly BIONETICS)*, ser. BICT'15. Brussels, BEL: ICST (Institute for Computer Sciences, Social-Informatics and Telecommunications Engineering) (2016), pp. 21–26 [Online]. https://doi.org/10.4108/eai.3-12-2015.2262516
12. N. Shone, T.N. Ngoc, V.D. Phai, Q. Shi, A deep learning approach to network intrusion detection. IEEE Trans. Emerg. Topics Comput. Intell. **2**(1), 41–50 (2018)
13. X. Li, W. Chen, Q. Zhang, L. Wu, Building auto-encoder intrusion detection system based on random forest feature selection. Comput. Secur. **95**, 101851 (2020) [Online]. http://www.sciencedirect.com/science/article/pii/S0167404820301231
14. L.R. Parker, P.D. Yoo, A.T. Asyhari, L. Chermak, Y. Jhi, K. Taha, Demise: interpretable deep extraction and mutual information selection techniques for IoT intrusion detection, in *Proceedings of ARES '19* (2019)

15. Y. Yu, J. Long, Z. Cai, Network intrusion detection through stacking dilated convolutional autoencoders. Secur. Commun. Netw. 4184196:1–4184196:10 (2017)
16. S. Park, M. Kim, S. Lee, Anomaly detection for http using convolutional autoencoders. IEEE Access **6**, 70884–70901 (2018)
17. Y. Xiao, C. Xing, T. Zhang, Z. Zhao, An intrusion detection model based on feature reduction and convolutional neural networks. IEEE Access **7**, 42210–42219 (2019)
18. Q.P. Nguyen, K.W. Lim, D.M. Divakaran, K.H. Low, M.C. Chan, GEE: a gradient-based explainable variational autoencoder for network anomaly detection, in *2019 IEEE Conference on Communications and Network Security (CNS)* (2019), pp. 91–99
19. L. Vu, V.L. Cao, Q.U. Nguyen, D.N. Nguyen, D.T. Hoang, E. Dutkiewicz, Learning latent distribution for distinguishing network traffic in intrusion detection system, in *ICC 2019—2019 IEEE International Conference on Communications (ICC)* (2019), pp. 1–6
20. S.J. Lee, P.D. Yoo, A.T. Asyhari, Y. Jhi, L. Chermak, C.Y. Yeun, K. Taha, Impact: impersonation attack detection via edge computing using deep autoencoder and feature abstraction. IEEE Access **8**, 65520–65529 (2020)
21. Y. Yang, K. Zheng, B. Wu, Y. Yang, X. Wang, Network intrusion detection based on supervised adversarial variational auto-encoder with regularization. IEEE Access **8**, 42169–42184 (2020)
22. V.L. Cao, M. Nicolau, J. McDermott, A hybrid autoencoder and density estimation model for anomaly detection, in *Proceedings of PPSN* (2016)
23. B. Zong, Q. Song, M.R. Min, W. Cheng, C. Lumezanu, D. ki Cho, H. Chen, Deep autoencoding Gaussian mixture model for unsupervised anomaly detection, in *Proceedings of ICLR* (2018)
24. C. Ieracitano, A. Adeel, F.C. Morabito, A. Hussain, A novel statistical analysis and autoencoder driven intelligent intrusion detection approach. Neurocomputing **387**, 51–62 (2020) [Online], http://www.sciencedirect.com/science/article/pii/S0925231219315759
25. I.J. Goodfellow, J. Pouget-Abadie, M. Mirza, B. Xu, D. Warde-Farley, S. Ozair, A.C. Courville, Y. Bengio, Generative adversarial nets, in *Proceedings of NIPS* (2014)
26. S. Shin, I. Lee, C. Choi, Anomaly dataset augmentation using the sequence generative models, in *2019 18th IEEE International Conference On Machine Learning And Applications (ICMLA)* (2019), pp. 1143–1148
27. B. Dowoo, Y. Jung, C. Choi, PcapGAN: packet capture file generator by style-based generative adversarial networks, in *2019 18th IEEE International Conference on Machine Learning and Applications (ICMLA)* (2019), pp. 1149–1154
28. L. Han, Y. Sheng, X. Zeng, A packet-length-adjustable attention model based on bytes embedding using flow-WGAN for smart cybersecurity. IEEE Access **7**, 82913–82926 (2019)
29. T. Schlegl, P. Seeböck, S. Waldstein, U. Schmidt-Erfurth, G. Langs, Unsupervised anomaly detection with generative adversarial networks to guide marker discovery (2017), pp. 146–157
30. H. Zenati, C.S. Foo, B. Lecouat, G. Manek, V.R. Chandrasekhar, Efficient GAN-based anomaly detection (2018). arXiv:1802.06222
31. T. Kohonen, The self-organizing map. Proc. IEEE **78**, 1464–1480 (1990)
32. H. Gunes Kayacik, A. Nur Zincir-Heywood, M.I. Heywood, A hierarchical SOM-based intrusion detection system. Eng. Appl. Artif. Intell. **20**(4), 439–451 (2007)
33. A. Ortiz, E. Hoz, E. De la Hoz, J. Ortega, B. Prieto, PCA filtering and probabilistic SOM for network intrusion detection. Neurocomputing **9** (2014)
34. O. Depren, M. Topallar, E. Anarim, M. Ciliz, An intelligent intrusion detection system (ids) for anomaly and misuse detection in computer networks. Expert Syst. Appl. **29**, 713–722 (2005)
35. M. Bahrololum, M. Khaleghi, Anomaly intrusion detection system using Gaussian mixture model, in *Proceedings of ICCIT* (2008), pp. 1162–1167
36. S. Parsazad, E. Saboori, A. Allahyar, Fast feature reduction in intrusion detection datasets, in *Proceedings of MIPRO* (2012), pp. 1023–1029
37. P. Casas, J. Mazel, P. Owezarski, Unsupervised network intrusion detection systems: detecting the unknown without knowledge. Comput. Commun. **35**, 772–783 (2012)
38. W.-C. Lin, S.-W. Ke, C.-F. Tsai, CANN: an intrusion detection system based on combining cluster centers and nearest neighbors. Knowl. Based Syst. **78**, 01 (2015)

39. W. Meng, W. Li, L.-F. Kwok, Design of intelligent KNN-based alarm filter using knowledge-based alert verification in intrusion detection. Secur. Commun. Netw. **8**(18), 3883–3895 (2015)
40. S. Mukherjee, N. Sharma, Intrusion detection using Naive Bayes classifier with feature reduction. Procedia Technol. **4**, 119–128 (2012)
41. D.M. Farid, M.Z. Rahman, Learning intrusion detection based on adaptive Bayesian algorithm, in *Proceedings of ICCIT* (2008), pp. 652–656
42. M. Albayati, B. Issac, Analysis of intelligent classifiers and enhancing the detection accuracy for intrusion detection system. Int. J. Comput. Intell. Syst. **8**, 841–853 (2015)
43. L. Koc, T.A. Mazzuchi, S. Sarkani, A network intrusion detection system based on a hidden Naïve Bayes multiclass classifier. Expert Syst. Appl. **39**(18), 13492–13500 (2012) [Online]. https://doi.org/10.1016/j.eswa.2012.07.009
44. Y. Wahba, E. ElSalamouny, G. ElTaweel, Improving the performance of multi-class intrusion detection systems using feature reduction (2015). arXiv:1507.06692
45. D. Barbara, N. Wu, S. Jajodia, Detecting novel network intrusions using Bayes (2001)
46. S.R. Safavian, D. Landgrebe, A survey of decision tree classifier methodology. IEEE Trans. Syst. Man Cybern. **21**(3), 660–674 (1991)
47. P.-F. Marteau, Sequence covering for efficient host-based intrusion detection. IEEE Trans. Inf. Forens. Secur. **14**, 994–1006 (2019)
48. H.G. Kayacik, A.N. Zincir-Heywood, M.I. Heywood, Selecting features for intrusion detection: a feature relevance analysis on KDD 99, in *Proceedings of PST* (2005)
49. C. Xiang, P.C. Yong, L.S. Meng, Design of multiple-level hybrid classifier for intrusion detection system using Bayesian clustering and decision trees. Pattern Recogn. Lett. **29**(7), 918–924 (2008) [Online]. https://doi.org/10.1016/j.patrec.2008.01.008
50. H.-J. Zimmermann, *Fuzzy Set Theory—and Its Applications*, 3rd edn. (Kluwer Academic Publishers, 1996)
51. A. Tajbakhsh, M. Rahmati, A. Mirzaei, Intrusion detection using Fuzzy association rules. Appl. Soft Comput. **9**(2), 462–469 (2009)
52. A.H. Hamamoto, L.F. Carvalho, L.D.H. Sampaio, T. Abro, M.L. Proena, Network anomaly detection system using genetic algorithm and fuzzy logic. Expert Syst. Appl. **92**(C), 390–402 (2018)
53. S. Elhag, A. Fernández, A. Altalhi, S. Alshomrani, F. Herrera, A multi-objective evolutionary fuzzy system to obtain a broad and accurate set of solutions in intrusion detection systems. Soft Comput. **23**(4), 1321–1336 (2019)
54. S. Kamalanathan, M. Karuppiah, S. Lakshmanan, S.H. Islam, M. Hassan, G. Fortino, K.-K.R. Choo, Intelligent temporal classification and fuzzy rough set-based feature selection algorithm for intrusion detection system in WSNs. Inform. Sci. **497**, 05 (2019)
55. J. Liu, Z. Wuxia, Z. Tang, Y. Xie, T. Ma, J. Zhang, G. Zhang, J. Niyoyita, Adaptive intrusion detection via GA-GOGMM-based pattern learning with fuzzy rough set-based attribute selection. Expert Syst. Appl. **139**, 112845 (2019)
56. G. Wang, J. Hao, J. Ma, L. Huang, A new approach to intrusion detection using artificial neural networks and fuzzy clustering. Expert Syst. Appl. **37**(9), 6225–6232 (2010)
57. S. Haykin, *Neural Networks: A Comprehensive Foundation*, 2nd edn. (Prentice Hall PTR, 1998)
58. F. Rosenblatt, Principles of neurodynamics: Perceptrons and the theory of brain mechanisms. Spartan Books, Washington DC (1961)
59. W. Tian, J. Liu, A new network intrusion detection identification model research, in *Proceedings of CAR*, vol. 2 (2010), pp. 9–12
60. Y. Yao, Y. Wei, F. Gao, Y. Yu, Anomaly intrusion detection approach using hybrid MLP/CNN neural network, in *Sixth International Conference on Intelligent Systems Design and Applications*, vol. 2 (2006), pp. 1095–1102
61. A. Saied, R.E. Overill, T. Radzik, Detection of known and unknown DDoS attacks using artificial neural networks. Neurocomputing **172**, 385–393 (2016)
62. C. Cortes, V. Vapnik, Support-vector networks. Mach. Learn. **20**(3), 273–297 (1995)

63. X. Bao, T. Xu, H. Hou, Network intrusion detection based on support vector machine, in *Proceedings of MASS* (2009), pp. 1–4
64. K. Zheng, X. Qian, P. Wang, Dimension reduction in intrusion detection using manifold learning, in *Proceedings of CIS*, vol. 2 (2009), pp. 464–468
65. B.-J. Kim, I.K. Kim, Kernel based intrusion detection system (2005), pp. 13– 18
66. G. Xiaoqing, G. Hebin, C. Luyi, Network intrusion detection method based on agent and SVM, in *Proceedings of ICIME* (2010), pp. 399–402
67. Y. Li, J. Xia, S. Zhang, J. Yan, X. Ai, K. Dai, An efficient intrusion detection system based on support vector machines and gradually feature removal method. Expert Syst. Appl. **39**, 424–430 (2012)
68. L. Breiman, Random forests. Mach. Learn. **45**(1), 5–32 (2001)
69. J. Zhang, M. Zulkernine, A. Haque, Random-forests-based network intrusion detection systems. IEEE Trans. Syst. Man Cybern. C Appl. Rev. **38**(5), 649–659 (2008)
70. P.-F. Marteau, S. Soheily-Khah, N. Béchet, Hybrid isolation forest—application to intrusion detection (2017). arXiv:1705.03800
71. A. Tesfahun, D.L. Bhaskari, Intrusion detection using random forests classifier with SMOTE and feature reduction, in *Proceedings of CUBE*, Nov 2013, pp. 127–132
72. R. Elbasiony, E.A. Sallam, T.E. Eltobely, M.M. Fahmy, A hybrid network intrusion detection framework based on random forests and weighted k-means (2013)
73. J. Li, Z. Zhao, R. Li, Machine learning-based IDS for software-defined 5G network. IET Netw. **7**(2), 53–60 (2018)
74. A. Madbouly, A. Gody, T. Barakat, Relevant feature selection model using data mining for intrusion detection system. Int. J. Eng. Trends Technol. **9**, 03 (2014)
75. S. Aljawarneh, M. Aldwairi, M. Yasin, Anomaly-based intrusion detection system through feature selection analysis and building hybrid efficient model. J. Comput. Sci. **25**, 03 (2017)
76. W. Hu, W. Hu, S. Maybank, Adaboost-based algorithm for network intrusion detection. IEEE Trans. Syst. Man Cybern. Part B (Cybernetics) **38**(2), 577–583 (2008)
77. S. Chebrolu, A. Abraham, J.P. Thomas, Feature deduction and ensemble design of intrusion detection systems. Comput. Secur. **24**(4), 295–307 (2005)
78. M. Gudadhe, P. Prasad, L. Kapil Wankhade, A new data mining based network intrusion detection model, in *Proceedings of ICCCT* (2010), pp. 731–735
79. H. Saxena, V. Richariya, Intrusion detection in KDD99 dataset using SVM-PSO and feature reduction with information gain. Int. J. Comput. Appl. **98**, 25–29 (2014)
80. Y. Gong, S. Mabu, C. Chen, Y. Wang, K. Hirasawa, Intrusion detection system combining misuse detection and anomaly detection using genetic network programming, in *Proceedings of ICCAS-SICE* (2009), pp. 3463–3467
81. R. Elhefnawy, H. Abounaser, A. Badr, A hybrid nested genetic-fuzzy algorithm framework for intrusion detection and attacks. IEEE Access **8**, 98218–98233 (2020)
82. Y. LeCun, Y. Bengio, G. Hinton, Deep learning. Nature **521**, 436–444 (2015)
83. Z. Li, Z. Qin, K. Huang, X. Yang, S. Ye, Intrusion detection using convolutional neural networks for representation learning, in *Proceedings of ICONIP* (2017)
84. M. Kalash, M. Rochan, N. Mohammed, N.D.B. Bruce, Y. Wang, F. Iqbal, Malware classification with deep convolutional neural networks, in *2018 9th IFIP International Conference on New Technologies, Mobility and Security (NTMS)* (2018), pp. 1–5
85. T. Kim, S.C. Suh, H. Kim, J. Kim, J. Kim, An encoding technique for CNN-based network anomaly detection, in *IEEE International Conference on Big Data (Big Data)* (2018), pp. 2960–2965
86. R. Blanco, P. Malagón, J. J. Cilla, J.M. Moya, Multiclass network attack classifier using CNN tuned with genetic algorithms, in *28th International Symposium on Power and Timing Modeling. Optimization and Simulation (PATMOS)* (2018), pp. 177–182
87. K. Wu, Z. Chen, W. Li, A novel intrusion detection model for a massive network using convolutional neural networks. IEEE Access **6**, 50850–50859 (2018)
88. S.Z. Lin, Y. Shi, Z. Xue, Character-level intrusion detection based on convolutional neural networks, in *International Joint Conference on Neural Networks (IJCNN)* (2018), pp. 1–8

89. L. Nie, Z. Ning, X. Wang, X. Hu, Y. Li, J. Cheng, Data-driven intrusion detection for intelligent internet of vehicles: a deep convolutional neural network-based method. IEEE Trans. Netw. Sci. Eng. 1 (2020)
90. S.-N. Nguyen, V.-Q. Nguyen, J. Choi, K. Kim, Design and implementation of intrusion detection system using convolutional neural network for dos detection, in *Proceedings of the 2nd International Conference on Machine Learning and Soft Computing*, ser. ICMLSC '18 (Association for Computing Machinery, New York, NY, USA, 2018), pp. 34–38 [Online]. https://doi.org/10.1145/3184066.3184089
91. Y. Jia, F. Zhong, A. Alrawais, B. Gong, X. Cheng, Flowguard: an intelligent edge defense mechanism against IoI DDoS attacks. IEEE Internet Things J. 1 (2020)
92. D.E. Rumelhart, G.E. Hinton, R.J. Williams, *Learning Internal Representations by Error Propagation* (MIT Press, Cambridge, MA, USA, 1986), pp. 318–362
93. H. Liu, B. Lang, M. Liu, H. Yan, CNN and RNN based payload classification methods for attack detection. Knowl. Based Syst. **163** (2018)
94. S. Lv, J. J. Wang, Y. Yang, J. Liu, Intrusion prediction with system-call sequence-to-sequence model. IEEE Access **6**, 71413–71421 (2018)
95. S. Hochreiter, J. Schmidhuber, Long short-term memory. Neural Comput. **9**, 1735–1780 (1997)
96. G. Kim, H. Yi, J. Lee, Y. Paek, S. Yoon, LSTM-based system-call language modeling and robust ensemble method for designing host-based intrusion detection systems (2016). arXiv:1611.01726
97. F. Jiang, Y. Fu, B.B. Gupta, Y. Liang, S. Rho, F. Lou, F. Meng, Z. Tian, Deep learning based multi-channel intelligent attack detection for data security. IEEE Trans. Sustain. Comput. **5**(2), 204–212 (2020)
98. O. Alkadi, N. Moustafa, B. Turnbull, K.R. Choo, A deep blockchain framework-enabled collaborative intrusion detection for protecting IoT and cloud networks. IEEE Internet Things J. 1 (2020)
99. R. Dong, X. Li, Q. Zhang, H. Yuan, Network intrusion detection model based on multivariate correlation analysis—long short-time memory network. IET Inf. Secur. **14**(2), 166–174 (2020)
100. R. Vinayakumar, K.P. Soman, P. Poornachandran, Applying convolutional neural network for network intrusion detection, in *2017 International Conference on Advances in Computing, Communications and Informatics (ICACCI)* (2017), pp. 1222–1228
101. W. Wang, Y. Sheng, J. Wang, X. Zeng, X. Ye, Y. Huang, M. Zhu, HAST-IDS: learning hierarchical spatial-temporal features using deep neural networks to improve intrusion detection. IEEE Access **6**, 1792–1806 (2018)
102. K. Jiang, W. Wang, A. Wang, H. Wu, Network intrusion detection combined hybrid sampling with deep hierarchical network. IEEE Access **8**, 32464–32476 (2020)
103. G.E. Hinton, Training products of experts by minimizing contrastive divergence. Neural Comput. **14**(8), 1771–1800 (2002) [Online]. https://doi.org/10.1162/089976602760128018
104. P. Smolensky, Chapter 6: Information processing in dynamical systems: foundations of harmony theory, in *Parallel Distributed Processing: Explorations in the Microstructure of Cognition*, ed. by D. E. Rumelhart, J. L. McLelland, vol 2 (Foundations, MIT Press, 1986), pp. 194–281
105. K. Alrawashdeh, C. Purdy, Reducing calculation requirements in FPGA implementation of deep learning algorithms for online anomaly intrusion detection, in *IEEE National Aerospace and Electronics Conference (NAECON)* (2017), pp. 57–62
106. S. Seo, S. Park, J. Kim, Improvement of network intrusion detection accuracy by using restricted Boltzmann machine, in *2016 8th International Conference on Computational Intelligence and Communication Networks (CICN)* (2016), pp. 413–417
107. N.T. Van, T.N. Thinh, L.T. Sach, An anomaly-based network intrusion detection system using deep learning, in *International Conference on System Science and Engineering (ICSSE)* (2017), pp. 210–214
108. M.Z. Alom, T.M. Taha, Network intrusion detection for cyber security using unsupervised deep learning approaches, in *IEEE National Aerospace and Electronics Conference (NAECON)* (2017), pp. 63–69

109. R. Salakhutdinov, G. Hinton, Deep Boltzmann machines, in *Proceedings of AISTATS 2009*, vol. 5 (2009), pp. 448–455
110. M.A. Salama, H.F. Eid, R.A. Ramadan, A. Darwish, A.E. Hassanien, Hybrid intelligent intrusion detection scheme (2011)
111. M.Z. Alom, V. Bontupalli, T.M. Taha, Intrusion detection using deep belief networks, in *National Aerospace and Electronics Conference (NAECON)* (2015), pp. 339–344
112. F. Qu, J. Zhang, Z. Shao, S. Qi, An intrusion detection model based on deep belief network, in *Proceedings of the 2017 VI International Conference on Network, Communication and Computing*, ser. ICNCC 2017 (Association for Computing Machinery, New York, NY, USA, 2017), pp. 97–101 [Online]. https://doi.org/10.1145/3171592.3171598
113. D. Liang, P. Pan, Research on intrusion detection based on improved DBN-ELM, in *2019 International Conference on Communications, Information System and Computer Engineering (CISCE)* (2019), pp. 495–499
114. G. Zhao, C. Zhang, L. Zheng, Intrusion detection using deep belief network and probabilistic neural network, in *2017 IEEE International Conference on Computational Science and Engineering (CSE) and IEEE International Conference on Embedded and Ubiquitous Computing (EUC)*, vol. 1 (2017), pp. 639–642
115. K. Alrawashdeh, C. Purdy, Toward an online anomaly intrusion detection system based on deep learning, in *2016 15th IEEE International Conference on Machine Learning and Applications (ICMLA)* (2016), pp. 195–200
116. N. Gao, L. Gao, Q. Gao, H. Wang, An intrusion detection model based on deep belief networks, in *Second International Conference on Advanced Cloud and Big Data* (2014), pp. 247–252
117. R.S. Sutton, A.G. Barto, *Introduction to Reinforcement Learning*, 1st edn. (MIT Press, Cambridge, MA, USA, 1998)
118. N. Sengupta, J. Sen, J. Sil, M. Saha, Designing of on line intrusion detection system using rough set theory and q-learning algorithm. Neurocomputing **111**, 161–168 (2013)

Part IV
Machine Learning/Deep Learning in Time Series Forecasting

Chapter 8
A Comparison of Contemporary Methods on Univariate Time Series Forecasting

Aikaterini Karanikola, Charalampos M. Liapis, and Sotiris Kotsiantis

Abstract In data science, time series forecasting is the process of utilizing past or present (known) observations of a target variable to make predictions about future (unknown) observations. Due to the usefulness of forecasting applications in numerous real-life problems, various Statistical and Machine Learning forecasting models have been proposed over recent years. The purpose of this chapter is to compare the performance of several contemporary forecasting models that are considered state of the art. These include Autoregressive Integrated Moving Average (ARIMA), Neural Basis Expansion Analysis (NBEATS), Probabilistic Time Series Modeling focusing on deep learning-based models and others. In the first section of this work a brief theoretical background of the methods is provided. Then, the experimental procedure is being described. For the comparison, 40 univariate time series of financial data that cover a 1-year period were used. A python repository of automated time series forecasting models (AtsPy) was exploited to run the experiments. For the final comparison three different metrics (RMSE, MAE and MAPE) were taken into consideration. The results of this extended experimental procedure are presented through various explanatory diagrams of the methods' performance in the final section.

8.1 Introduction

The history of forecasting is rooted back in ancient times. Back then, predicting the future was a sign of divine enlightenment, although it was not uncommon for fortunetellers to be accused of possessing unholy powers. Pythia, the high priestess

A. Karanikola · C. M. Liapis · S. Kotsiantis (✉)
University of Patras, Rion, Greece
e-mail: sotos@math.upatras.gr

A. Karanikola
e-mail: karanikola@upatras.gr

C. M. Liapis
e-mail: c.liapis@upnet.gr

G. A. Tsihrintzis et al. (eds.), *Advances in Machine Learning/Deep Learning-based Technologies*, Learning and Analytics in Intelligent Systems 23,
https://doi.org/10.1007/978-3-030-76794-5_8

of Apollo in the sanctuary of Delphi in Greece, is one of the most famous examples of ancient forecasters. She was supposed to speak Apollo's will and was consulted for numerous decisions throughout classical years, and, even if they were often ambiguous, her oracles played an important role in the outcome of many historical events. Although the list of the enlighted priests, prophets, and clairvoyants through centuries is long, the profound, mainly with the scientific revolution, transformation of scientific practices in the acquisition of human knowledge and the subsequent continuous evolution of technology closed these stories in the realm of fantasy, approaching the human need for predictions about the future from a radically different perspective.

Today forecasting is a science subfield, consisted of statistical and machine learning models able to predict the future, or, to be more specific, to make predictions about the values of the variables that are used to describe the examined problems. The goal remains the same. If we can guess how the various parameters of a phenomenon will evolve, then we can make better decisions about it, minimizing any emerging risks. Consequently, the field of time series forecasting is rapidly growing in recent years, as the need for accurate forecasting models is increasing continuously. These days forecasting models are being used in various domains, like predicting the weather [34], the rainfall rate [53], stocks' closing values [65], communications [72], earthquakes [5], even the evolution of diseases' outbreaks, like the recent COVID-19 pandemic [39, 47].

Data is the keystone element in every machine learning concept. Before discussing different time series forecasting approaches, it is necessary to describe the form of the data in time series forecasting problems, or, in other words, to define what a time series is, as well as to mention some essential concepts related to it. A time series is a set of numerical observations, each of which is being recorded at a specific timestamp, which is used as an index. When observations are being recorded in fixed time intervals, then the time series is referred to as discrete-time time series, otherwise, when observations are recorded continuously in a given time interval, it is referred to as continuous-time. Considering the number of variables that vary over time, time series are grouped in two categories [66]: univariate, where only one variable varies over time, and multivariate, where more than one variable vary simultaneously over time. Time series are often depicted in simple 2-D graphs in which every observation is marked by a point. The horizontal axis represents time, while the vertical axis represents magnitude whose behavior is being studied over time. In Fig. 8.1 the closing price (in USD) for the Microsoft Corporation stock can be seen from January 2018 to December 2018, forming a time series consisted of 365 observations.

Time series are formed by four different components, which can be decomposed from the recorded data. These are referred to as trend, seasonal variations, cyclic variations and irregular movements [1, 31, 55] and they will be briefly explained below.

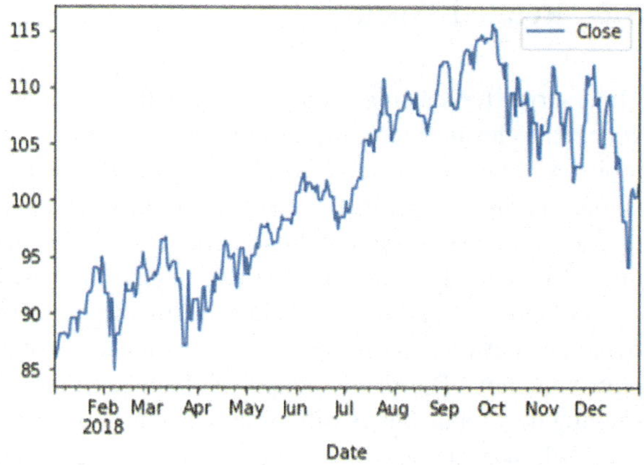

Fig. 8.1 Closing value of MSFT stock during 2018. Observations had been sampled daily over a period of 1 year

- **Trend:** Trend is a long-term component that depicts the general tendency of a time series to continuously increase (upward trend) or decrease (downward trend). Its pattern may be linear or non-linear and may remain fixed or change over time.
- **Seasonality:** Seasonal variations are regular, relatively short-term repetitive up-and-down fluctuations that, unlike cycles below, occur within a specific period, for example a week, a month, or a year.
- **Cycles:** Cyclical variations are long-term, up-and-down, and potentially irregular swings which are caused by circumstances that occur in cycles with no standard duration.
- **Irregularities:** Irregular are the variations that do not follow an expected pattern and are caused by unpredictable events. Here we can include any variation that cannot be categorized as trend, cyclical, or seasonal.

Taking into account the four elements discussed above, the models exploiting these attributes in a decompositional strategy are typically additive or multiplicative [31]. In additive models, every observation is a sum of the four aforementioned components, implying that they are independent, while in multiplicative methods observations are formed by the multiplication of the elements, assuming that they may affect each other. The additive model is preferred when the magnitude of the trend, cycle, and seasonal components remain constant over the course of the series. Multiplicative methods are preferred when the magnitude of the components varies as the level of the trend varies. In real-world problems often all kinds of components may or may not be present. Time series analysis identifies and isolates the components, to explore the way each one of them affects the observed data, and thus to make future predictions possible by fitting the correct model. Actually, only the first three components form stable patterns in time series, while irregular variations act like random errors. A forecast is the act of extending these stable patterns to the future, based on the idea that what happened in the past is possible to reoccur.

8.2 Related Work

Time series forecasting, being a very active field of research in recent years, has very rich relevant literature. In this section, aspects of previous work related to the topic presented in this study will be briefly addressed, focusing particularly on recent studies about time series forecasting on financial data, as well as on studies seeking to draw conclusions about which methods outperform others in specific tasks, given that such general scope conclusions can be established.

Economy, in addition to being an important factor that greatly influences the dynamics between countries, constitutes the key pillar of the prosperity of contemporary society. The development of the stock market has played a special role in shaping new financial conditions and the entry of a growing number of companies in the stock market has significantly changed the economic activity of recent decades. Predicting the behavior of the stock market has become particularly important for stakeholders, companies and investors. Many well-known times series forecasting methods have been used for this reason. Autoregressive Integrated Moving Average (ARIMA) models have been especially popular in the stock market's time series prediction [17, 63]. Furthermore, hybrid methods exploiting similar approaches have also been introduced. In [70] a new method that combines ARIMA with Empirical Wavelet Transform (EWT) technique along with improved Artificial Bee Colony (ABC) algorithm and Extreme Learning Machine (ELM) neural network is introduced. An ARIMA combination with Long Short-Term Memory (LSTM) Recurrent Neural Networks (RNN) has been presented in [14], while ARIMA combined with Back Propagation Neural Network (BPNN) model has been proposed in [69] to predict the Chinese stock market.

Additionally, exponential smoothing models are also common in predicting financial times series. Single and double exponential smoothing models were used for this purpose in [57], while Brown's double exponential smoothing method was used in [26]. Different variants of the Holt-Winters additive model (HWAM), which is an extension of Holt's exponential smoothing, has also been exploited on financial data in [44]. A hybrid methodology between Empirical Mode Decomposition with the Holt-Winters method (EMD-HW) was used to improve forecasting performances in financial time series in [3], while EDM-HW bagging was presented in [4].

Machine Learning based forecasting methods have also drawn attention in recent years and have been extensively used in financial problems [58, 59]. Vector Autoregressive Models and Recurrent Neural Networks (RNN) have been used [6], as well as Autoregressive Neural Networks for predicting stock returns in [50]. A hybrid method combining RNNs with ARMA and exponential smoothing was presented in [51], while the exploitation of Deep learning networks has been presented in [15]. Convolutional Neural Networks (CNNs) [8] as well as Deep CNNs [13] have also been used for stock market predictions. Deep learning approaches have led to more sophisticated methods, like the one in [20] where CNNs are combined with LSTM stacked autoencoders (SAE) to forecast univariate time series.

Prophet, Facebook's automatic forecasting procedure, has recently been used, among others, in bitcoin prize prediction [19], the forecasting of COVID-19 daily cases in Bangladesh [38] and sales forecasting [73].

Given that this work focuses mainly on comparing different algorithms for predicting time series and to draw useful conclusions especially in terms of financial data covering a relatively short period of time (1 year), it would be a great omission not to refer to other studies aimed at similar comparisons. This 1-year specification was chosen in order to compare the reviewed methods, which in this work are employed in an automated fashion, under somewhat adverse conditions. Such conditions, where past data have a rather small extent, may and have occurred throughout the history[1] and the behavior of our available tools in such a setting needs, in our view, to be investigated.

Much of past work focuses on comparing ARIMA models with other approaches, as ARIMA models are among the most common methods for predicting time series. They include, among others, ARIMA comparisons with LSTM [56] over univariate financial data, with the Holt-Winters model over primary energy consumption data [49], with Artificial Neural Network (ANN) over medical supplies data [42] and traffic data [36], with ANN hybrid methods over financial data [41] and with BPNN while forecasting the price of bitcoin [12]. As for methods related to Artificial Neural Networks, a variety of Neural Network models have been compared with the Box-Jenkins and the Holt-Winters methods in [21], while in [48] various ANN architectures are being compared using stock market's data. Moreover, a comparison of RNNs including an empirical study using both LSTM and GRU networks is presented in [71], while CNNs are tested against a standard backpropagation NN with one hidden layer and LSTM networks in [35]. An interesting comparison between ANN, hybrid methods, and multiple regression techniques is presented in [74].

This work aims to fill the gap that exists in the parallel comparison of multiple methods representing different approaches. Although similar studies have been conducted offering important feedback about the compared methods, usually the comparison involves only a few approaches. Studies comparing multiple approaches cover only a small proportion of the literature. Such works include a six method, specifically Moving Average (MA), Multivariate adaptive regression splines (MARS), ARIMA, Grey Model (GM), ANN and SVM, comparison in [10], ANN, Bayesian Models and Stochastic approaches comparisons in [64], and lastly [46], in which different machine learning based approaches are being compared using univariate time series consisted of weather data

[1] Actually, COVID-19 outbreak is in fact such a scenario, the consequences of which can be seen, among others, in various financial data.

8.3 Theoretical Background

In this section the essential theoretical elements of the algorithms reviewed will be outlined. These include the following 6 classes, in a broad sense, of methods: ARIMA [24], Facebook's Prophet [62], the Holt-Winters Seasonal models, meaning the Additive (HWAAS) and Multiplicative (HWAMS) models [11], N-Beats [45], a DeepAR implementation from the Gluonts python library [2, 54] and finally a series of Trigonometric BATS methods from [37].

8.3.1 ARIMA

ARIMA(p, d, q) [24] stands for Auto-Regressive Integrated Moving Average. These models are a generalization, applied in non-stationary cases, of the Auto Regressive Moving Average (ARMA) [1] models which are fitted to time series data for analyzing and forecasting. The added Integrated part refers to Differencing, a method of transforming a non-stationary time series into a stationary one.

In an Autoregression (AR) model the time-dependent variable is forecast as a linear combination of its past values, as a regression, in other words, of the variable against itself. With p we denote the model's lag order, meaning the number of lag observations—previous values which are included in the linear combination. The manipulation of $\phi_1, \dots, \phi_p$ results in different time series patterns. Thus, given the coefficients and the error-noise, it is

$$y_t = c + \phi_1 y_{t-1} + \phi_2 y_{t-2} + \cdots + \phi_p y_{t-p} + \varepsilon_t, \tag{8.1}$$

In a similar manner, a Moving Average (MA) model forecasts with a regression-like logic where, instead of using the past values themselves, the model uses past forecast errors. Given that q is the order of moving average and that, as above, changing $\theta_1, \dots, \theta_q$ different time series patterns occur, the formulation is:

$$y_t = c + \varepsilon_t + \theta_1 \varepsilon_{t-1} + \theta_2 \varepsilon_{t-2} + \cdots + \theta_q \varepsilon_{t-q}. \tag{8.2}$$

Lastly, the transformation via differencing of a non-stationary time series into a stationary one, given the differencing order d, is being accomplished by subtracting an observation from an observation at the previous time step. The degree d denotes the number of times the initial observations are being subtracted and it is chosen so that it is sufficiently large to transform our initial time series into a stationary one.

8.3.2 Prophet

Prophet is a modular regression model. It has interpretable parameters that can be intuitively adjusted by analysts with domain knowledge about the time series [62]. It is essentially a decomposable additive model which consists of three main components: trend, seasonality, and holidays. Given the noise, these components are simply combined, similarly to a Generalized Additive Model (GAM) [28] with the use of time as a regressor,[2] in the formalization as follows:

$$y(t) = g(t) + s(t) + h(t) + \varepsilon_t \tag{8.3}$$

where $g(t)$ models non-periodic changes, $s(t)$ represents periodic changes like daily, weekly or yearly seasonality, $h(t)$ represents effects on a potentially irregular schedule over a day or a period of days and finally ε_t is every other change not modeled by the main components.

Prophet utilizes two different trend models. The first one is a saturating growth model, in which growth is nonlinear and saturates at a carrying capacity. This kind of growth is described by the basic logistic growth model, modified in a way to adapt to the fact that both growth rate and carrying capacity are not constant. The second is a piece-wise linear model, which is utilized on problems that do not exhibit saturating growth. To fit the effects of seasonality Prophet relies on Fourier series, making the algorithm flexible, while the Fourier order defines the kind of seasonality (yearly, weekly, etc.). As holidays and events can lead to—fortunately predictable—shocks to business time series, Prophet is equipped with a list of such events, tackling the problem of these periodically occurring incidents by incorporating them into the forecast. Concluding, it is also crucial for the method's efficiency to be fitted in time series with strong seasonalities and several seasons of past data.

8.3.3 The Holt-Winters Seasonal Models

There are two variations in the Holt-Winters' seasonal method:

- The Holt-Winters' additive method (HWAAS): Exponential Smoothing with additive trend and additive seasonality
- The Holt-Winters' multiplicative method (HWAMS): Exponential Smoothing with additive trend and multiplicative seasonality.

[2] Also with linear and non-linear time functions as components [62].

8.3.3.1 HWAAS

The Holt-Winters Additive Model [30, 68] is an extension to Holt's method that models seasonality by adding such an additional factor to the trended forecast. The formula has three components each of which is an equation with a corresponding smoothing parameter. Specifically:

$$\hat{y}_{t+h|t} = \ell_t + hb_t + s_{t+h-m(k+1)} \tag{8.4}$$

$$\ell_t = \alpha(y_t - s_{t-m}) + (1-\alpha)(\ell_{t-1} + b_{t-1}) \tag{8.5}$$

$$b_t = \beta^*(\ell_t - \ell_{t-1}) + (1-\beta^*)b_{t-1} \tag{8.6}$$

$$s_t = \gamma(y_t - \ell_{t-1} - b_{t-1}) + (1-\gamma)s_{t-m} \tag{8.7}$$

where, α, β, γ are the smoothing factors, ℓ_t is the level, b_t the trend and s_t seasonal component.

8.3.3.2 HWAMS

As mentioned, the HWAAS and HWAMS are essentially the same method in two versions. The two variations differ in the seasonal component's form. The Multiplicative Seasonality version incorporates the exact same logic as before with the below differences, as seen in the formalization:

$$\hat{y}_{t+h|t} = (\ell_t + hb_t)s_{t+h-m(k+1)} \tag{8.8}$$

$$\ell_t = \alpha\frac{y_t}{s_{t-m}} + (1-\alpha)(\ell_{t-1} + b_{t-1}) \tag{8.9}$$

$$b_t = \beta^*(\ell_t - \ell_{t-1}) + (1-\beta^*)b_{t-1} \tag{8.10}$$

$$s_t = \gamma\frac{y_t}{(\ell_{t-1} + b_{t-1})} + (1-\gamma)s_{t-m} \tag{8.11}$$

8.3.4 N-BEATS: Neural Basis Expansion Analysis

N-Beats is a deep neural architecture for time series forecasting. The architecture is based on backward and forward residual links and a very deep stack of fully-connected layers [45]. Assuming the desired output size is H, the input data size of the time series will be an integer multiple m of H. The mH dimensional input goes through Stacks of multiple basic Blocks, a trend and a seasonality stack. Every basic Block is comprised of a 4-layer fully connected stack, before being divided into two parts, each of which leads through another FC. The one part is a future point prediction of length H, and the other an input data prediction of length mH. Every Stack is arranged in a Double (Backcast and Forecast) Residual Stacking topology. In each of them, the input to every consecutive Block is a mH dimensional vector, which is the result of an element-wise subtraction performed between the previous Block's Backcast output and input. The output of a Stack consists of a mH dimensional Stack Backcast Output which is fed into the next Stack and a H dimensional Stack Forecast Output, which will be summed element-wise together with the respective outputs from every Stack to form the final H dimensional Forecast output vector. To sum up, the overall architecture is comprised of two stacks arranged so that the trend stack is followed by the seasonality stack [45], along with a double residual stacking topology combined with the aforementioned forecast–backcast principle.

8.3.5 DeepAR

DeepAR is a deep learning method for estimating the probability distribution of a time series' future given its past. The DeepAR algorithm is realized by the implementation of an RNN model similar to the one described in [54]. The DeepAR model is based on an autoregressive recurrent network architecture [25, 54, 61] specifically utilizing an LSTM-based architecture [29], trained on a large number of related time series in order to tackle the problem of producing accurate probabilistic forecasts. The model uses such an LSTM architecture which enables the simultaneous training of related time-series and exploits an encoder–decoder setup, common in sequence-to-sequence models, where both the encoder and decoder networks share the same architecture. The suggestion is to create global models from related time series using RNN architecture with a Gaussian/Negative Binomial likelihood in combination with nonlinear data transformations. Concluding, DeepAR also implements Monte Carlo sampling and does not require extended feature engineering, as it exploits given covariates to learn seasonal behavior. The overall strategy is aiming to outcome probabilistic forecasts that outperform traditional single-item forecasting.

8.3.6 Trigonometric BATS

Lastly, moving on to the final of our brief presentations of the models used, which consists of a series of methods that combine in their fundamental components different statistical approaches, all of which have in common a trend element and the use of a Trigonometric seasonal formulation [27] and can be applied to a wide range of time-series problems [37]. The first of these various modifications is TATS. The latter is an identifier for the algorithm's basic components which the method utilizes using a Trigonometric Seasonal formulation [27], enabling the decomposition of complex seasonal time series and modeling non-integer seasonal frequencies, an ARMA errors section [1] and trend modeling component. The TBAT adds a Box–Cox transformation [7] manipulating non-normal dependent variables and handling non-linearity, maintaining the trigonometric component and leaving out a seasonal.

Finally, following the aforementioned, TBATS consists of: Trigonometric, Box–Cox transform, ARMA errors, Trend, and Seasonal components.

8.4 Experiments and Results

Moving on to the presentation of the experimental procedure and the consequent results. The section consists of four parts, each of which contains respectively information about the data used and the aforementioned algorithms, the evaluation metrics and finally explanatory tables and graphs of the results.

8.4.1 Datasets

For the experimental procedure 40 datasets drawn from *Yahoo! Finance* webpage were used. *Yahoo! Finance* was chosen with the intention of having the desired data publicly available, so that the results could be easily recreated. To form the final datasets, 40 of the most active stocks were chosen. Each stock is denoted by a characteristic abbreviation of its name. The full name of each stock and its corresponding abbreviation can be seen in Table 8.1. For every stock, the formed dataset is covering a 1 year period from 1/4/2018 to 31/3/2019. Given that this work is focused only on univariate time series forecasting, each dataset consists of the closing market value of the stock daily, and, as a result, contains about 365 values. To ensure that the acquired datasets do not contain any missing data, considering that not all the algorithms can handle missing values, for the days that the stock market remained closed the closing value of the previous day was used.

8.4.2 Algorithms

As has already been emphasized, the core of this work is the comparison of several widely used, as well as powerful, algorithms for univariate time series forecasting, which are considered state of the art. As for the algorithms' implementation, the AtsPy [60] framework was used, which is a recently developed repository of automated structural and machine learning time series models in Python. The AtsPy repository was chosen for it includes a variety of automated time series forecasting models, while at the same time can reduce the structural model errors of existing singular models about 30–50% by using a Gradient Boosting Model (GBM) [22, 40] with time-series extracted features. The repository uses the already existing python libraries *statsmodels*, *fbprophet*, *gluonts*, *pmdarima* and *tbats* to exploit all the aforementioned in Sect. 8.3 algorithms in a unified way, and aspires to constitute an easy to use, yet effective automated tool for time series forecasting. In addition to the above,

Table 8.1 List of the stocks that were used in experimental procedure

No	Ab.	Name	No	Ab.	Name
1	AAL	American Airlines Group Inc.	21	INO	Inovio Pharmaceuticals, Inc.
2	AAPL	Apple Inc.	22	INTC	Intel Corporation
3	AMD	Advanced Micro Devices, Inc.	23	LUV	Southwest Airlines Co.
4	APA	Apache Corporation	24	MSFT	Microsoft Corporation
5	AUY	Yamana Gold Inc.	25	MU	Micron Technology, Inc.
6	BA	The Boeing Company	26	NCLH	Norwegian Cruise Line Holdings Ltd.
7	BABA	Alibaba Group Holding Limited	27	ORCL	Oracle Corporation
8	BAC	Bank of America Corporation	28	PE	Parsley Energy, Inc.
9	BB	BlackBerry Limited	29	PG	The Procter & Gamble Company
10	CCL	Carnival Corporation & Plc	30	PYPL	PayPal Holdings, Inc.
11	CSCO	Cisco Systems, Inc.	31	SBUX	Starbucks Corporation
12	DAL	Delta Air Lines, Inc.	32	SIRI	Sirius XM Holdings Inc.
13	DIS	The Walt Disney Company	33	SNAP	Snap Inc.
14	EBAY	eBay Inc.	34	SPOT	Spotify Technology S.A.
15	ERI	Eldorado Resorts, Inc.	35	TGS	Transportadora de Gas del Sur S.A.
16	F	Ford Motor Company	36	TSLA	Tesla, Inc.
17	FB	Facebook, Inc.	37	TWTR	Twitter, Inc.
18	GM	General Motors Company	38	UAL	United Airlines Holdings, Inc.
19	HPE	Hewlett Packard Enterprise Company	39	WFC	Wells Fargo & Company
20	HPQ	HP Inc.	40	ZNGA	Zynga Inc.

it is also stated that it automatically identifies seasonalities in data by using spectrum analysis, periodograms, and pick analysis. Up until now, as can be easily derived from the foregoing, AtsPy provides implementations of the following algorithms, the basic elements of which are summarized below[3]:

- **ARIMA(p,d,q):** Auto-Regressive Integrated Moving Average, which is an evolution of the AutoRegressive Moving Average (ARMA). The arguments p, d, and q define respectively the number of time lags of the autoregressive model (p), the degree of differencing (d), and the order of the moving-average model (q).
- **Prophet:** Modeling Multiple Seasonality With Linear or Non-linear Growth. A procedure based on an additive model where non-linear trends are fit—in our case—with daily seasonality.
- **HWAAS:** Exponential Smoothing with Additive Trend and Additive Seasonality.
- **HWAMS:** Exponential Smoothing with Additive Trend and Multiplicative Seasonality.
- **NBEATS:** A deep neural architecture based on backward and forward residual links and a very deep stack of fully connected layers.
- **Gluonts [DeepAR]:** GluonTS is a python library for deep-learning-based time series modeling [2]. Here a Recurrent Neural Network (RNN) based model exploiting the benefits of deep learning in time series forecasting is used.
- **TBAT:** Method based on Trigonometric Seasonal, ARMA error and Trend component that incorporates Box–Cox transformations.
- **TATS:** Method based on Trigonometric Seasonal, ARMA error, Seasonal and Trend components, without using Box–Cox transformations.
- **TBATS1:** Method similar to TBAT which additionally exploits one seasonal period.
- **TBATS2:** Method similar to TBATS1 which exploits two seasonal periods instead of one.
- **TBATP1:** Method similar to TBATS1 where seasonal inference is hardcoded by periodicity.

During the experimental procedure, every algorithm was tested on each one of the 40 datasets. Considering ARIMA(p,d,q) the parameter order that defines the values of p,d and q was set to $(p, d, q) = (0, 1, 0)$. For Gluonts [DeepAR] and NBEATS the number of epochs was set to 20. To execute HWAAS and HWAMS, the model of exponential smoothing was tuned by setting the parameter *trend* equal to `'add'` for HWASS and to `'mul'` for HWAMS, to apply additive and multiplicative trend respectively. The parameters `seasonal_period` and `using_box_cox` of the TBATS python implementation were tuned properly to perform TBAT, TATS, TBATS1, TBATP1 and TBATS2 each time. Moreover, unless stated otherwise, all parameters were set to default.

In the course of the algorithms' execution, two different kinds of predictions took place: an in-sample and an out-sample prediction. For the in-sample prediction the

---

[3] All of the implemented in AtsPy framework algorithms were used during the whole experimental procedure.

datasets were split into two subsets. The first one consisted of 70%—the first 254 instances—of the dataset, and the second one consisted of the remaining 30%—the last 111 instances. The first subset was used as a training test to predict the values of the remaining 30% of the dataset, while the values of the second subset were not revealed to the model, but used as a control set to make assumptions for the performance of the compared methods. For the out-sample prediction the whole dataset was used to train the model and the algorithm was requested to predict the values of the target variable for two weeks following the last date of the training test. Although there are methods to draw conclusions from both types of prediction, in this work we focus on the results based on the in-sample one, as they tend to be more reliable.

8.4.3 Evaluation

To compare the algorithms, multiple evaluation metrics were used. As there is not a universal metric that can lead to overall safe results, since all of them have their advantages and disadvantages [9, 32, 67], three different evaluation metrics that are widely used in the evaluation of forecasting techniques were utilized. These are: Mean Absolute Error (MAE), Root Mean Squared Error (RMSE), and Mean Absolute Percentage Error (MAPE). Assuming that y is the actual value of an observation, while $\hat{y}$ is the predicted one, a brief explanation of how these metrics are computed along with some useful information, is provided below:

- **Mean Absolute Error** (MAE): MAE is the arithmetic average of absolute errors, and is computed by the analytical expression:

$$MAE(y, \hat{y}) = \frac{1}{n}\sum_{i=1}^{n} \left|y_i - \hat{y}_i\right| \tag{8.12}$$

  MAE is a scale-dependent accuracy measure, so it cannot be used while comparing series with different scales but it is commonly used for comparing different models when data are homogenous. It is an easy to interpret measure and robust to outliers.
- **Root Mean Squared Error** (RMSE): RMSE is the squared root of the average of the squares of the forecast errors. It is computed by the following analytical formula:

$$RMSE(y, \hat{y}) = \sqrt{\frac{1}{n}\sum_{i=1}^{n}(y_i - \hat{y}_i)} \tag{8.13}$$

  RMSE is also scale-dependent and as a result, it is not suitable for comparison between different datasets but it is appropriate for use when comparing several models over the same dataset. RMSE is highly sensitive to the existence of outliers, as large errors can increase dramatically its value due to the squared difference in its formula. RMSE is non-negative and in general lower RMSE values indicate better performance, while a value equal or almost equal to zero suggests perfect forecast.

- **Mean Absolute Percentage Error** (MAPE): MAPE is the average of the absolute percentage errors computed by the formula:

$$MAPE(y, \hat{y}) = \left\{ \frac{1}{n} \sum_{i=1}^{n} \left| \frac{y_i - \hat{y}_i}{y_i} \right| \right\} \times 100 \tag{8.14}$$

MAPE is one of the most common metrics used in the comparison of forecasting methods as is easily interpretable, as well as not scale-dependent, so it can be used not only to compare methods used in different datasets [43]. MAPE is considered as a percentage error, but also can score values above 100 when the predicted value is extremely lower or higher than the actual one. Another drawback is that MAPE is not defined when the actual value is equal to zero or could lead to extreme values when the actual ones tend to zero. In this work, not such values occurred, thus MAPE is trustworthy to use.

8.4.4 Results

In this section, the results of the experimental procedure along with the results of the corresponding statistical tests are presented. As already mentioned, the performance of 11 methods over 40 discrete stock market datasets was compared using three different metrics (MAE, RMSE, and MAPE). Due to the number of methods and datasets that were used, the exact results of the experimental procedure are depicted in quite spacious tables. Considering the limitations of the available space and the fact that large tables with numeric results are usually more frustrating than explanatory, the complete tables of results for the three metrics are not provided here, but more efficient ways to present the results of this work were chosen, such as wins/ties/losses (w/t/l) tables, cd diagrams, and box-plots. For everyone interested the complete tables of the results of all three evaluation metrics can be found at: https://rb.gy/7lqayb.

To ensure that the results of the whole procedure are trustworthy and to examine the statistical independence of the compared forecasting algorithms, the experiments were followed by statistical tests. For this reason, non-parametric all versus all Friedman test [23] was used, along with Bonferroni–Dunn post-hoc test [18], setting the parameter a equal to 0.05. A Friedman test accepts or rejects the null hypothesis H_0 that the means of the results of two or more algorithms are the same, and if the null hypothesis is rejected, then all the methods or groups of them are not similar. Moreover, in this case, the Bonferroni–Dunn post-hoc test detects the groups of methods that differ. The statistical tests not only provide information about which of the algorithms perform similarly and which do not, but in addition the Friedman test ranks the tested methods by a value, which indicates the algorithms that outperform their rivals.

Given that three different metrics were used as evaluation measures, the rest of this section is split into three parts, each one dedicated to MAE, RMSE, and MAPE respectively. The following are provided for each one of the aforementioned metrics: (a) a win/ties/losses table, presenting the number of wins, ties, and losses of each pair of methods, (b) a table featuring the Friedman Ranking score, (c) a cd-diagram of the compared methods and (d) a box-plot for graphic summarization of the results of the presented metric.

The w/t/l tables provide information on how many times one algorithm beats another, considering all the possible pairs. In each (i,j) cell of the table lies a triplet of numbers, which indicate the number of wins, ties, and losses considering the pair of algorithms in the i-th row and j-th column. As a result, the w/t/l tables are always symmetric with respect to the diagonal, while all the values on the diagonal are equal to zero. In this work, a win or a loss is considered to have occurred in the cases where the difference between the performances of the algorithm in the under consideration metric is greater than 0.01, otherwise it is considered a tie. In Friedman ranking tables, the algorithms are ranked during the statistical test, from the best to worst, considering the examined measure. The lower the value assigned by the Friedman test, the higher the achieved performance. Considering the cd diagrams, the score of Friedman's ranking is used to compute the critical difference value. When the numerical difference of the ranking score between the two methods is greater than the critical difference value, the methods are considered statistically dependent. In cd diagrams, methods that are statistically dependent are connected with bold horizontal lines. Box-plots provide useful information about the median, the first and the third quartile, as well as the minimum and maximum values that the compared methods scored. Outliers were not plotted, since they tended to push and press the boxes over the edges of the plot, making it illegible.

8.4.4.1 MAE

According to the box-plot in Fig. 8.2 that depicts the performance of the MAE metric, NBEATS achieves lower scores than its rivals, followed by TATS, when Prophet scores the highest scores. The remaining algorithms have similar performance. The same conclusion is also drawn from the w/t/l table in Table 8.2, in which NBEATS manages to achieve more wins and beat all of its pairs. The statistical test that took place, rejected the null hypothesis H_0, as the p-value was equal to 0.00001, thus not all algorithms have similar performance. The groups of statistical dependent methods, that act in a similar way can be seen in Fig. 8.3, in the corresponding cd diagram. Friedman's ranking, that is presented in Table 8.3, provides us with a better perspective about the performance of the rest algorithms, sorting them in ascending order from NBEATS, which achieves the lowest MAE to the Prophet, which achieves the highest.

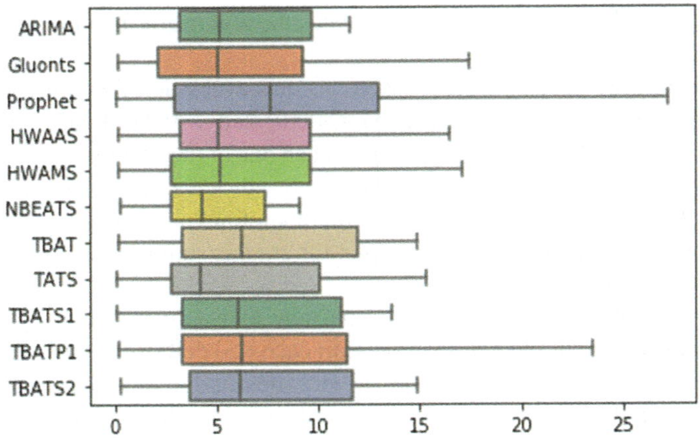

Fig. 8.2 Box-plot of MAE metric performance. Essential information about the distributional characteristics of the MAE scores concerning all 40 datasets, as well as the level of the MAE scores are presented

Table 8.2 W/T/L table of MAE metric

	ARIMA	Gluonts	Prophet	HWAAS	HWAMS	NBEATS	TBAT	TATS	TBATS1	TBATP1	TBATS2
ARIMA	0/0/0	19/1/20	10/0/30	18/0/22	17/0/23	27/0/13	11/2/27	23/3/14	17/1/22	13/3/24	10/0/30
Gluonts	20/1/19	0/0/0	15/0/25	21/0/19	22/0/18	22/0/18	17/1/22	23/0/17	19/0/21	19/1/20	19/0/21
Prophet	30/0/10	25/0/15	0/0/0	29/0/11	28/0/12	31/0/9	22/1/17	29/0/11	25/1/14	24/1/15	22/0/18
HWAAS	22/0/18	19/0/21	11/0/29	0/0/0	15/5/20	27/0/13	16/0/24	25/0/15	19/0/21	17/0/23	19/0/21
HWAMS	23/0/17	18/0/22	12/0/28	20/5/15	0/0/0	26/0/14	15/0/25	27/0/13	18/0/22	17/0/23	17/0/23
NBEATS	13/0/27	18/0/22	9/0/31	13/0/27	14/0/26	0/0/0	12/0/28	16/0/24	11/0/29	12/0/28	11/0/29
TBAT	27/2/11	22/1/17	17/1/22	24/0/16	25/0/15	28/0/12	0/0/0	25/9/6	11/24/5	9/28/3	8/26/6
TATS	14/3/23	17/0/23	11/0/29	15/0/25	13/0/27	24/0/16	6/9/25	0/0/0	10/8/22	8/9/23	5/8/27
TBATS1	22/1/17	21/0/19	14/1/25	21/0/19	22/0/18	29/0/11	5/24/11	22/8/10	0/0/0	9/19/12	7/18/15
TBATP1	24/3/13	20/1/19	15/1/24	23/0/17	23/0/17	28/0/12	3/28/9	23/9/8	12/19/9	0/0/0	7/18/15
TBATS2	30/0/10	21/0/19	18/0/22	21/0/19	23/0/17	29/0/11	6/26/8	27/8/5	15/18/7	15/18/7	0/0/0

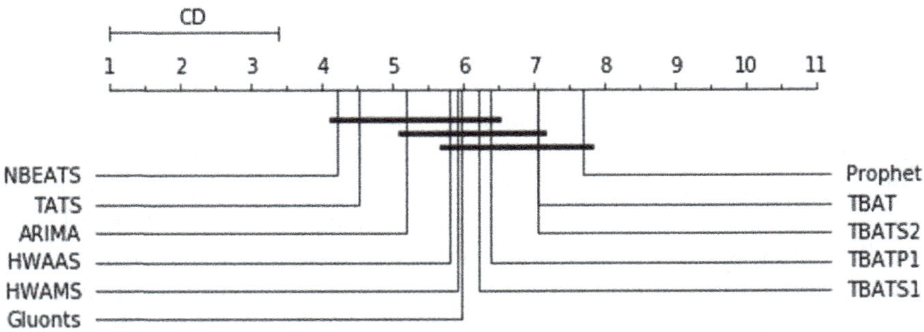

Fig. 8.3 CD-plot of MAE metric. Algorithms whose distance on the horizontal axis is less that the computed cd value have similar performance. Thus, they are statistical dependent and are connected with bold horizontal lines

Table 8.3 Friedman's ranking for MAE, RMSE, MAPE

	MAE		RMSE		MAPE	
1	NBEATS	4.225	TATS	4.375	NBEATS	3.15
2	TATS	4.525	NBEATS	4.6	Gluonts	4.7
3	ARIMA	5.2	ARIMA	5.15	TATS	5.3
4	HWAAS	5.8	HWAAS	5.85	HWAAS	5.475
5	HWAMS	5.9	Gluonts	5.875	ARIMA	5.55
6	Gluonts	5.975	HWAMS	5.9	HWAMS	5.625
7	TBATS1	6.2125	TBATS1	6.1625	TBATS1	6.4875
8	TBATP1	6.3875	TBATP1	6.4625	TBATP1	6.6125
9	TBATS2	7.0375	TBATS2	6.9625	TBATS2	7.3875
10	TBAT	7.0375	TBAT	7.0125	TBAT	7.5625
11	Prophet	7.7	Prophet	7.65	Prophet	8.15

8.4.4.2 RMSE

Furthermore, using RMSE as an evaluation metric similar results can be derived. The box-plot in Fig. 8.4 suggests that the top two performing algorithms are NBEATS and TATS. On the other side, it is clear that Prophet performs the worst, while the second-worst performance is achieved by TBAT. The w/t/l table in Table 8.4 provides justification for these assumptions, as TATS achieves the most wins followed by NBEATS. As for the statistical tests, the null hypothesis was rejected by a p-value

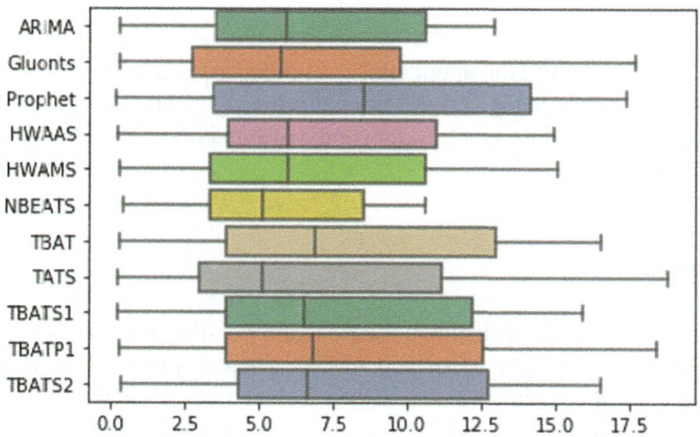

Fig. 8.4 Box-plot of RMSE metric. Essential information about the distributional characteristics of the RMSE scores concerning all 40 datasets, as well as the level of the RMSE scores are presented

Table 8.4 W/T/L table of RMSE metric

	ARIMA	Gluonts	Prophet	HWAAS	HWAMS	NBEATS	TBAT	TATS	TBATS1	TBATP1	TBATS2
ARIMA	0/0/0	19/1/20	12/0/28	16/2/22	16/0/24	22/0/18	11/2/27	24/2/14	18/1/21	13/3/24	10/1/29
Gluonts	20/1/19	0/0/0	14/0/26	21/0/19	22/0/18	22/0/18	18/0/22	22/0/18	19/0/21	18/0/22	18/0/22
Prophet	28/0/12	26/0/14	0/0/0	29/0/11	27/1/12	30/0/10	23/0/17	29/0/11	27/0/13	25/0/15	22/0/18
HWAAS	22/2/16	19/0/21	11/0/29	0/0/0	17/4/19	27/0/13	16/0/24	26/0/14	19/0/21	18/0/22	18/0/22
HWAMS	24/0/16	18/0/22	12/1/27	19/4/17	0/0/0	25/0/15	15/1/24	27/0/13	18/1/21	18/1/21	16/0/24
NBEATS	18/0/22	18/0/22	10/0/30	13/0/27	15/0/25	0/0/0	13/0/27	19/0/21	13/0/27	13/0/27	12/0/28
TBAT	27/2/11	22/0/18	17/0/23	24/0/16	24/1/15	27/0/13	0/0/0	26/8/6	11/24/5	9/28/3	9/26/5
TATS	14/2/24	18/0/22	11/0/29	14/0/26	13/0/27	21/0/19	6/8/26	0/0/0	8/8/24	8/8/24	6/7/27
TBATS1	21/1/18	21/0/19	13/0/27	21/0/19	21/1/18	27/0/13	5/24/11	24/8/8	0/0/0	9/19/12	8/18/14
TBATP1	24/3/13	22/0/18	15/0/25	22/0/18	21/1/18	27/0/13	3/28/9	24/8/8	12/19/9	0/0/0	7/19/14
TBATS2	29/1/10	22/0/18	18/0/22	22/0/18	24/0/16	28/0/12	5/26/9	27/7/6	14/18/8	14/19/7	0/0/0

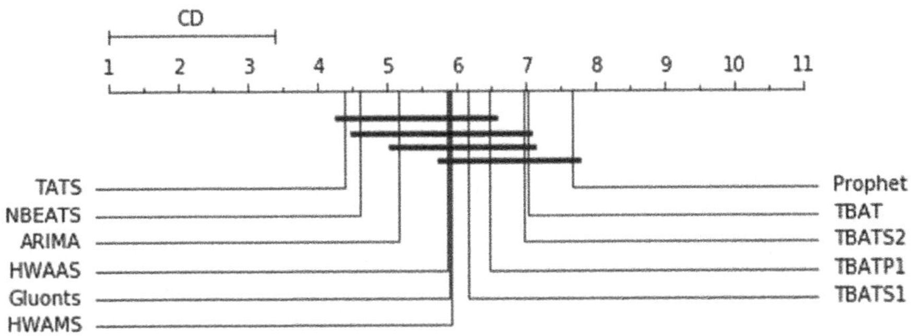

Fig. 8.5 CD-plot of RMSE metric. Statistical dependent algorithms are connected with bold horizontal lines, while algorithms that are not connected have statistically different behavior

equal to 0.00003. Information about statistical independence of the methods is presented in the cd diagram in Fig. 8.5. Additionally, the algorithms are ranked according to the Friedman test, whose results confirm the aforementioned assumptions can be found in Table 8.3.

8.4.4.3 MAPE

MAPE was the last of the calculated metrics, and possibly the most instructive, as it is not scale dependent like MAE and RMSE. The box plot of this metric, which can be found in Fig. 8.6, enables us to depict a clear view about the performance of the compared methods, as the differences between them become now conspicuous. NBEATS achieves the lowest percentage error, while the one that follows is Gluonts. Prophet has the highest error percentage. The same conclusions are also drawn from the w/t/l table in Table 8.5 and the results of Friedman's ranking in Table 8.3. According to the ranking by Friedman's test, the rank values of the mediocre rated methods

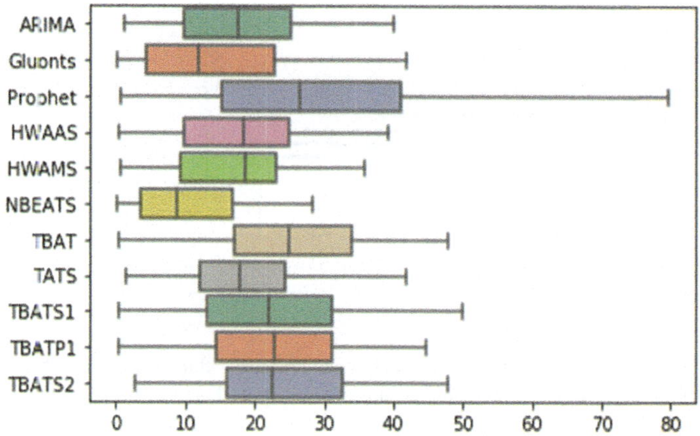

Fig. 8.6 Box-plot of MAPE metric performance. Essential information about the distributional characteristics of the MAPE scores concerning all 40 datasets, as well as the level of the MAPE scores are presented

Table 8.5 W/T/L table of MAPE metric

	ARIMA	Gluonts	Prophet	HWAAS	HWAMS	NBEATS	TBAT	TATS	TBATS1	TBATP1	TBATS2
ARIMA	0/0/0	24/0/16	11/0/29	18/1/21	20/0/20	28/0/12	12/0/28	23/0/17	17/0/23	16/1/23	11/0/29
Gluonts	16/0/24	0/0/0	9/0/31	18/0/22	20/0/20	22/0/18	11/0/29	16/0/24	13/0/27	12/0/28	11/0/29
Prophet	29/0/11	31/0/9	0/0/0	33/0/7	33/0/7	32/0/8	24/0/16	29/0/11	28/0/12	26/0/14	21/0/19
HWAAS	21/1/18	22/0/18	7/0/33	0/0/0	17/3/20	30/0/10	10/0/30	24/0/16	16/0/24	15/0/25	17/0/23
HWAMS	20/0/20	20/0/20	7/0/33	20/3/17	0/0/0	33/0/7	10/2/28	21/1/18	16/2/22	16/2/22	15/0/25
NBEATS	12/0/28	18/0/22	8/0/32	10/0/30	7/0/33	0/0/0	5/0/35	7/0/33	8/0/32	6/0/34	5/0/35
TBAT	28/0/12	29/0/11	16/0/24	30/0/10	28/2/10	35/0/5	0/0/0	23/8/9	11/24/5	10/27/3	9/25/6
TATS	17/0/23	24/0/16	11/0/29	16/0/24	18/1/21	33/0/7	9/8/23	0/0/0	11/7/22	12/8/20	8/7/25
TBATS1	23/0/17	27/0/13	12/0/28	24/0/16	22/2/16	32/0/8	5/24/11	22/7/11	0/0/0	10/19/11	7/18/15
TBATP1	23/1/16	28/0/12	14/0/26	25/0/15	22/2/16	34/0/6	3/27/10	20/8/12	11/19/10	0/0/0	7/18/15
TBATS2	29/0/11	29/0/11	19/0/21	23/0/17	25/0/15	35/0/5	6/25/9	25/7/8	15/18/7	15/18/7	0/0/0

are quite similar, in contrast to the ranking values of the best and worst algorithms which show a greater difference. In this case, the null hypothesis was rejected by a p-value equal to 0.00001. The corresponding cd diagram is presented in Fig. 8.7.

8.4.4.4 Comparisons

As has already been mentioned, MAE and RMSE metrics provide us with useful information about the performance of the compared methods, but, being both scale-dependent, they can not constitute a standalone criterion for reaching valid conclusions. MAPE was used to overcome this limitation. The comparison of these metrics' outcomes leads to more reliable interpretations on various aspects of the whole forecast procedure.

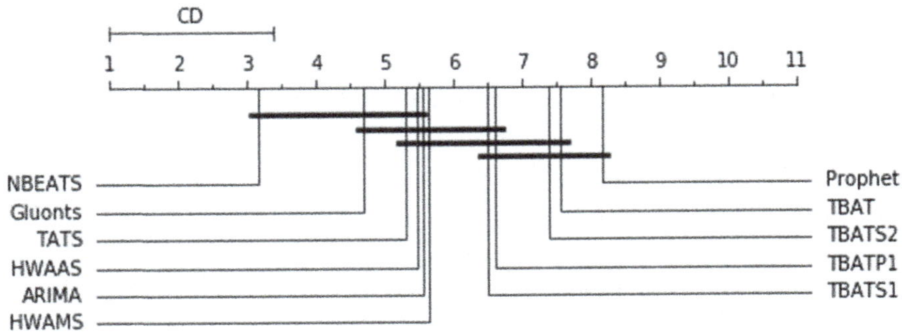

Fig. 8.7 CD-plot of MAPE metric. Statistical dependent algorithms concerning MAPE are connected with bold horizontal lines, forming groups of algorithms that have statistically similar behavior

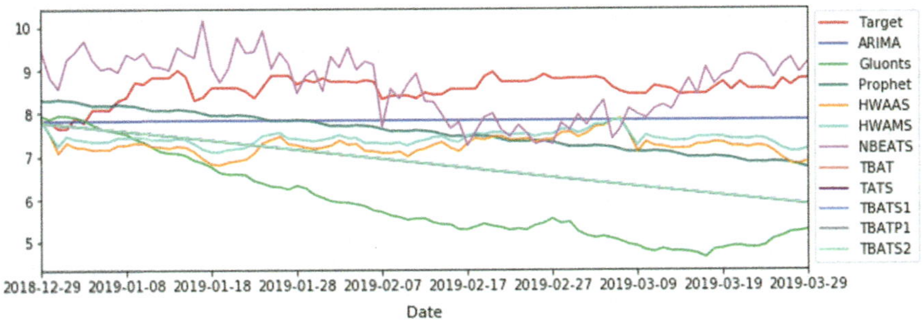

Fig. 8.8 In-sample forecast predictions for F dataset

In all three metrics, NBEATS was in the top two methods that scored the lowest errors. Specifically, NBEATS was ranked first considering MAE and MAPE and second considering RMSE. TATS also achieved remarkable—within the overall norms of these results—performances, ranked first in RMSE and second in MAE. A prima facie unexpected fact is that while Gluonts' DeepAR implementation was ranked sixth and fifth in MAE and RMSE respectively, it was also ranked second considering MAPE. ARIMA and HWAAS also performed fairly well, as in all the metrics they were both ranked between the third and the fifth place. Facebook's Prophet scores the lowest ranking in all metrics, which is quite expected, as, among others, the quality of its performance is often proportional to data covering longer periods of time. Finally, TBAT was ranked as the second from the end in all cases, while the other methods that belong to its family, meaning the TBATS1, TBATP1, and TBATS2, were also placed in seventh, eighth, and ninth position respectively. Indicative diagrams portraying the methods' forecasting performance concerning in-sample prediction are presented in Figs. 8.8, 8.9 and 8.10.

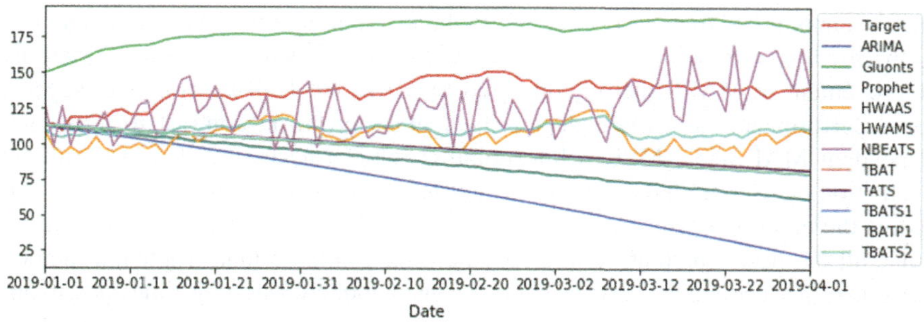

Fig. 8.9 In-sample forecast predictions for SPOT dataset

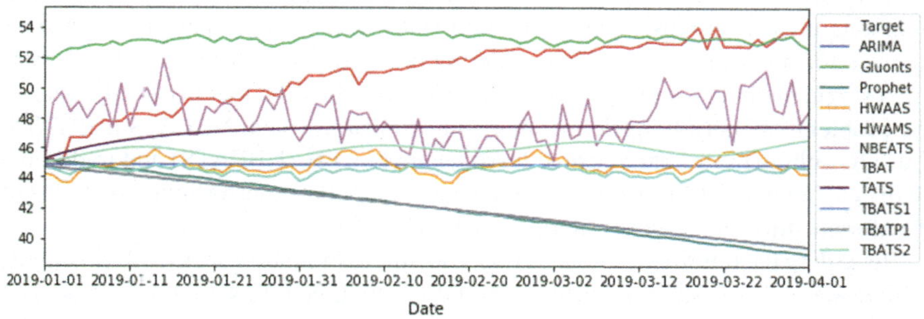

Fig. 8.10 In-sample forecast predictions for ORCL dataset

8.5 Conclusions

The purpose of this work was to compare the performance of different time series forecasting approaches over univariate time series that are relatively short. Eleven methods, belonging to six discrete families and emerging from the fields of statistics or machine learning, were utilized. The data were chosen to cover a 1-year period in order to examine the case where only a short amount of past observations are available for making predictions, given that this scenario is common in real-world tasks. However, financial data predictions, like forecasting the stock market closing values which were used in this study, being easily affected by a variety of factors, are characterized by a high level of difficulty. Moreover here, the experimental procedure was conducted in an automated fashion, a specification which was a fixed parameter of this study, as such a manipulation of the available data, given that it is reliable, can make predicting and conclusion-reaching less time-consuming and much more easily performed, always under the limitations of the inevitable trade-offs.

The non-existence of a universal metric for evaluating the performance of such forecasting methods pointed to the use of three different formulas (MAE, RMSE, MAPE) and the comparison of their respective scores as the concluding tool for reliable interpretations. NBEATS and TATS seem to dominate their rivals in all three metrics, making it into the top three of Friedman's ranking procedure. A notable fact is that python's implementation of DeepAR, Gluonts, exhibits significant ranking improvement in MAPE metric compared to MAE and RMSE. This may suggest good performance, as MAPE depicts percentage errors in contrast to the scale-dependent MAE and RMSE, while pointing out the necessity of using multiple evaluation metrics for having better insight in results. Another remark, which reinforces the empirical principle that there is no single method that can satisfactorily solve all the problems that lie in a specific task, would be that Prophet, a method capable of performing accurate predictions when applied over long-time data, in our case exhibits poor performance. Statistical tests also provided useful information about the obtained outcomes. Moreover, in all cases, the best and worst-performing methods were statistical independent, a fact that indicates valid comparisons. Cd-diagrams present groups of methods with statistically similar performance, once more confirming and somewhat emphasizing the above general remark, that there is not a single method—statistical or machine learning based—that fits in all forecasting problems. Finally, considering short-time univariate data, NBEATS and TATS, thus algorithms from both machine learning approaches (NBEATS) and traditional statistical methods (TATS), outperformed their competitors performing—in terms of the aforementioned metrics—similarly, achieving the most accurate results.

Regarding the future extensions of this work, the exploitation of ensemble methods of time-series forecasting algorithms should be examined. Ensembles are based on the idea that using a set of diverse models and combining their outputs, generally leads to better obtained performance than using a single model at a time [16, 52]. By the exploitation of the latter, the combination of statistical and machine learning methods thus maintaining the benefits of both approaches, may lead to the reduction of the produced forecasting errors. Concluding, the case of long-period data covering several years should be examined, both in terms of this chapter's specifications, as under the aforementioned ensemble strategy.

References

1. R. Adhikari, R.K. Agrawal, An Introductory Study on Time Series Modeling and Forecasting (2013). https://doi.org/10.1210/jc.2006-1327. arXiv Preprint arXiv:13026613
2. A. Alexandrov, K. Benidis, M. Bohlke-Schneider, V. Flunkert, J. Gasthaus, T. Januschowski, D.C. Maddix, S. Rangapuram, D. Salinas, J. Schulz, L. Stella, A.C. Türkmen, Y. Wang, GluonTS: probabilistic time series models in python. (2019), pp. 1–2
3. A.M. Awajan, M.T. Ismail, A hybrid approach EMD-HW for short-term forecasting of daily stock market time series data, in *AIP Conf Proc*, vol. 1870 (2017). https://doi.org/10.1063/1.4995933
4. A.M. Awajan, M.T. Ismail, Wadi S. Al, Improving forecasting accuracy for stock market data using emd-hw bagging. PLoS One **13**, 1–20 (2018). https://doi.org/10.1371/journal.pone.0199582
5. A. Barkat, A. Ali, U. Hayat, Q.G. Crowley, K. Rehman, N. Siddique, T. Haidar, T. Iqbal, Time series analysis of soil radon in Northern Pakistan: Implications for earthquake forecasting. Appl. Geochem. **97**, 197–208 (2018). https://doi.org/10.1016/j.apgeochem.2018.08.016

6. J.M. Binner, T. Elger, B. Nilsson, J.A. Tepper, Tools for Non-Linear Time Series Forecasting in Economics - an Empirical Comparison of Regime Switching Vector Autoregressive Models and Recurrent Neural Networks. Adv. Econ. **19**, 71–91 (2004). https://doi.org/10.1016/S0731-9053(04)19003-8
7. G.E.P. Box, D.R. Cox, An Analysis of Transformations. J. R. Stat. Soc. Ser. B **26**, 211–243 (1964)
8. J. Cao, J. Wang, Stock price forecasting model based on modified convolution neural network and financial time series analysis. Int. J. Commun. Syst. **32**, 1–13 (2019). https://doi.org/10.1002/dac.3987
9. T. Chai, R.R. Draxler, Root mean square error (RMSE) or mean absolute error (MAE)? Arguments against avoiding RMSE in the literature. Geosci. Model Dev. **7**, 1247–1250 (2014). https://doi.org/10.5194/gmd-7-1247-2014
10. H.K. Chan, S. Xu, X. Qi, A comparison of time series methods for forecasting container throughput. Int. J. Logist. Res. Appl. **22**, 294–303 (2019). https://doi.org/10.1080/13675567.2018.1525342
11. C. Chatfield, The Holt-Winters forecasting procedure. Appl. Stat. **27**, 264 (1978). https://doi.org/10.2307/2347162
12. C.C. Chen, J.H. Chang, F.C. Lin, J.C. Hung, C.S. Lin, Y.H. Wang, Comparison of forcasting ability between backpropagation network and ARIMA in the prediction of bitcoin price, in *Proc—2019 Int Symp Intell Signal Process Commun Syst ISPACS*, 2019–2020 (2019). https://doi.org/10.1109/ISPACS48206.2019.8986297
13. J.F. Chen, W.L. Chen, C.P. Huang, S.H. Huang, A.P. Chen, Financial time-series data analysis using deep convolutional neural networks, in *Proc—2016 7th Int Conf Cloud Comput Big Data. CCBD*, vol. 2016 (2017), pp. 87–92. https://doi.org/10.1109/CCBD.2016.027
14. H.K. Choi, Stock Price Correlation Coefficient Prediction with ARIMA-LSTM Hybrid Model (2018)
15. E. Chong, C. Han, F.C. Park, Deep learning networks for stock market analysis and prediction: methodology, data representations, and case studies. Expert. Syst. Appl. **83**, 187–205 (2017). https://doi.org/10.1016/j.eswa.2017.04.030
16. T.G. Dietterich, Ensemble methods in machine learning, in *mult classif syst*, vol. 1857(1–15) (2000), pp. 45014–45019. https://doi.org/10.1007/3-540-45014-9
17. Y. Dong, S. Li, X. Gong, Time series analysis: an application of Arima model in stock price forecasting, vol. 29 (2017), pp. 703–710. https://doi.org/10.2991/iemss-17.2017.140
18. O.J. Dunn, Multiple comparisons among means. J. Am. Stat. Assoc. **56**, 52 (1961). https://doi.org/10.2307/2282330
19. F. Duvodq, V.W.X. Nkdv, H.G.X. Wu, Dataset A ARIMA e ProphetFB. C:6–9 (2016)
20. A. Essien, C. Giannetti, A Deep learning framework for univariate time series prediction using convolutional LSTM stacked autoencoders, in *IEEE Int Symp Innov Intell Syst Appl INISTA 2019—Proc* (2019), pp. 1–6. https://doi.org/10.1109/INISTA.2019.8778417
21. J. Faraway, C. Chatfield, Time series forecasting with neural networks: a comparative study using the airline data. J. R. Stat. Soc. Ser. C Appl. Stat. **47**, 231–250 (1998). https://doi.org/10.1111/1467-9876.00109
22. J.H. Friedman, Greedy function approximation: a gradient boosting machine. Ann. Stat. **29**, 1189–1232 (2001). https://doi.org/10.1214/aos/1013203451
23. M. Friedman, The use of ranks to avoid the assumption of normality implicit in the analysis of variance. J. Am. Stat. Assoc. **32**, 675–701 (1937). https://doi.org/10.1080/01621459.1937.10503522
24. M. Geurts, G.E.P. Box, G.M. Jenkins, Time series analysis: forecasting and control. J. Mark. Res. **14**, 269 (1977). https://doi.org/10.2307/3150485
25. A. Graves, *Generating Sequences with Recurrent Neural Networks* (2013), pp. 1–43
26. S. Hansun, A new approach of Brown's double exponential smoothing method in time series analysis. Balk. J. Electr. Comput. Eng. **4**, 75–78 (2016). https://doi.org/10.17694/bajece.14351
27. A. Harvey, S.J. Koopman, M. Riani, The modeling and seasonal adjustment of weekly observations. J. Bus. Econ. Stat. **15**, 354–368 (1997). https://doi.org/10.1080/07350015.1997.10524713

28. T. Hastie, R. Tibshirani, Generalized additive models: some applications. J. Am. Stat. Assoc. **82**, 371–386 (1987). https://doi.org/10.1080/01621459.1987.10478440
29. S. Hochreiter, LSTM can solve hard long time lag problems, in *Adv Neural Inf Process Syst* (1997), pp. 473–479
30. C.C. Holt, Forecasting seasonals and trends by exponentially weighted moving averages. Int. J. Forecast. **20**, 5–10 (2004). https://doi.org/10.1016/j.ijforecast.2003.09.015
31. R.J. Hyndman, G. Athanasopoulos, Forecasting: principles and practice, in Principles of Optimal Design (2018)
32. R.J. Hyndman, A.B. Koehler, Another look at measures of forecast accuracy. Int. J. Forecast. **22**, 679–688 (2006). https://doi.org/10.1016/j.ijforecast.2006.03.001
33. A.T. Jebb, L. Tay, W. Wang, Q. Huang, Time series analysis for psychological research: examining and forecasting change. Front. Psychol. **6**, 1–24 (2015). https://doi.org/10.3389/fpsyg.2015.00727
34. Z. Karevan, J.A.K. Suykens, Transductive LSTM for time-series prediction: an application to weather forecasting. Neural Netw. **125**, 1–9 (2020). https://doi.org/10.1016/j.neunet.2019.12.030
35. I. Koprinska, D. Wu, Z. Wang, Convolutional neural networks for energy time series forecasting, in *Proc Int Jt Conf Neural Networks 2018* (2018). https://doi.org/10.1109/IJCNN.2018.8489399
36. L. Lenferink, A comparison between artificial neural networks and ARIMA models in traffic forecasting (2019), pp. 1–12
37. A.M. De Livera, R.J. Hyndman, R.D. Snyder, Forecasting time series with complex seasonal patterns using exponential smoothing. Monash Univ Work Pap to appear (2010). https://doi.org/10.1198/jasa.2011.tm09771
38. S. Mahmud, Bangladesh COVID-19 daily cases time series analysis using facebook prophet model (2020). https://doi.org/10.13140/RG.2.2.23220.68481
39. M. Maleki, M.R. Mahmoudi, D. Wraith, K.H. Pho, Time series modelling to forecast the confirmed and recovered cases of COVID-19. Travel Med. Infect. Dis. **101742** (2020). https://doi.org/10.1016/j.tmaid.2020.101742
40. L. Mason, J. Baxter, P. Bartlett, M. Frean, Boosting algorithms as gradient descent in function space. Nips (1999). 10.1109/5.58323
41. N. Merh, V.P. Saxena, K.R. Pardasani, A comparison between hybrid approaches of ANN and Arima for indian stock trend forecasting. Bus. Intell. J. **3**, 23–43 (2010)
42. A. Molina, B. Ponte, J. Parreno, D. De la Fuente, J. Costas, Forecasting erratic demand of medicines in a public hospital: a comparison of artificial neural networks and ARIMA models, in *Proc 2016 Int Conf Artif Intell ICAI 2016—WORLDCOMP 2016* (2016), pp. 401–406
43. A. de Myttenaere, B. Golden, B. Le Grand, F. Rossi, Mean absolute percentage error for regression models. Neurocomputing **192**, 38–48 (2016). https://doi.org/10.1016/j.neucom.2015.12.114
44. M.M. Navarro, B.B. Navarro, Optimal short-term forecasting using GA-based Holt-Winters method (2019), pp. 681–685
45. B.N. Oreshkin, D. Carpov, N. Chapados, Y. Bengio, N-BEATS: Neural basis expansion analysis for interpretable time series forecasting (2019), pp. 1–31
46. G. Papacharalampous, H. Tyralis, D. Koutsoyiannis, Univariate time series forecasting of temperature and precipitation with a focus on machine learning algorithms: a multiple-case study from Greece. Water Resour. Manag. **32**, 5207–5239 (2018). https://doi.org/10.1007/s11269-018-2155-6
47. V. Papastefanopoulos, P. Linardatos, S. Kotsiantis, COVID-19: a comparison of time series methods to forecast percentage of active cases per population. Appl. Sci. **10**, 1–15 (2020). https://doi.org/10.3390/app10113880
48. L. Di. Persio, O. Honchar, Artificial neural networks architectures for stock price prediction: comparisons and applications. Int. J. Circuits Syst. Signal Process. **10**, 403–413 (2016)
49. A. Rahman, A.S. Ahmar, Forecasting of primary energy consumption data in the United States: a comparison between ARIMA and Holter-Winters models, in *AIP Conf Proc*, vol. 1885 (2017). https://doi.org/10.1063/1.5002357

50. A.M. Rather, A prediction based approach for stock returns using autoregressive neural networks, in *Proc 2011 World Congr Inf Commun Technol WICT 2011*. (2011), pp. 1271–1275. https://doi.org/10.1109/WICT.2011.6141431
51. A.M. Rather, A. Agarwal, V.N. Sastry, Recurrent neural network and a hybrid model for prediction of stock returns. Expert. Syst. Appl. **42**, 3234–3241 (2015). https://doi.org/10.1016/j.eswa.2014.12.003
52. O. Sagi, L. Rokach, Ensemble learning: a survey, in *Wiley Interdiscip Rev Data Min Knowl Discov*, vol. 8 (2018), pp. 1–18. https://doi.org/10.1002/widm.1249
53. S.Q. Salih, A. Sharafati, I. Ebtehaj, H. Sanikhani, R. Siddique, R.C. Deo, H. Bonakdari, S. Shahid, Z.M. Yaseen, Integrative stochastic model standardization with genetic algorithm for rainfall pattern forecasting in tropical and semiarid environments. Hydrol. Sci. J. **65**, 1145–1157 (2020). https://doi.org/10.1080/02626667.2020.1734813
54. D. Salinas, V. Flunkert, J. Gasthaus, T. Januschowski, DeepAR: probabilistic forecasting with autoregressive recurrent networks. Int. J. Forecast. (2019). https://doi.org/10.1016/j.ijforecast.2019.07.001
55. L. Seymour, P.J. Brockwell, R.A. Davis, *Introduction to Time Series and Forecasting*. (1997)
56. S. Siami-Namini, N. Tavakoli, A. Siami Namin, A comparison of ARIMA and LSTM in forecasting time series, in *Proc—17th IEEE Int Conf Mach Learn Appl ICMLA 2018* (2019), pp. 1394–1401. https://doi.org/10.1109/ICMLA.2018.00227
57. F. Sidqi, I.D. Sumitra, Forecasting product selling using single exponential smoothing and double exponential smoothing methods, in *IOP Conf Ser Mater Sci Eng*, vol. 662. (2019). https://doi.org/10.1088/1757-899X/662/3/032031
58. R. Singh, S. Srivastava, Stock prediction using deep learning. Multimed. Tools Appl. **76**, 18569–18584 (2017). https://doi.org/10.1007/s11042-016-4159-7
59. N. Sirimevan, I.G.U.H. Mamalgaha, C. Jayasekara, Y.S. Mayuran, C. Jayawardena, Stock market prediction using machine learning techniques, in *Int. Conf Adv Comput ICAC*, pp. 192–197 (2019). https://doi.org/10.1109/ICAC49085.2019.9103381
60. D. Snow, AtsPy: Automated Time Series Models in Python (1.15). (2020)
61. I. Sutskever, O. Vinyals, Q.V. Le, Sequence to sequence learning with neural networks. Adv. Neural Inf. Process. Syst. **4**, 3104–3112 (2014)
62. S.J. Taylor, B. Letham, Business time series forecasting at scale. PeerJ Prepr. 5e3190v2 **35**, 48–90 (2017). https://doi.org/10.7287/peerj.preprints.3190v2
63. S.A.L. Wadi, M. Almasarweh, A.A. Alsaraireh, Predicting closed price time series data using ARIMA model. Mod. Appl. Sci. **12**, 181 (2018). https://doi.org/10.5539/mas.v12n11p181
64. M.B. Wagena, D. Goering, A.S. Collick, E. Bock, D.R. Fuka, A. Buda, Z.M. Easton, Comparison of short-term streamflow forecasting using stochastic time series, neural networks, process-based, and Bayesian models. Environ. Model. Softw. **126**, 104669 (2020). https://doi.org/10.1016/j.envsoft.2020.104669
65. J. Wang, J. Wang, Neurocomputing forecasting stock market indexes using principle component analysis and stochastic time effective neural networks. Neurocomputing **156**, 68–78 (2015). https://doi.org/10.1016/j.neucom.2014.12.084
66. W.W.S. Wei, *Time Series Analysis Univariate and Multivariate Methods*. (Pearson Education, 2018)
67. C.J. Willmott, K. Matsuura, Advantages of the mean absolute error (MAE) over the root mean square error (RMSE) in assessing average model performance. Clim. Res. **30**, 79–82 (2005). https://doi.org/10.3354/cr030079
68. P.R. Winters, Forecasting sales by exponentially weighted moving averages. Manag. Sci. **6**, 324–342 (1960). https://doi.org/10.1287/mnsc.6.3.324
69. L. Xiong, Y. Lu, Hybrid ARIMA-BPNN model for time series prediction of the Chinese stock market, in *2017 3rd Int Conf Inf Manag ICIM 2017*, pp. 93–97 (2017). https://doi.org/10.1109/INFOMAN.2017.7950353
70. H. Yu, L.J. Ming, R. Sumei, Z. Shuping, A hybrid model for financial time series forecasting-integration of EWT, ARIMA with the improved ABC optimized ELM. IEEE Access. **8**, 84501–84518 (2020). https://doi.org/10.1109/ACCESS.2020.2987547

71. G.P. Zhang, Neural networks for time-series forecasting, in *Handbook of National Computing*, vol. 1–4, (2012). pp. 461–477. https://doi.org/10.1007/978-3-540-92910-9_14
72. K. Zhang, G. Chuai, W. Gao, X. Liu, S. Maimaiti, Z. Si, A new method for traffic forecasting in urban wireless communication network. EURASIP J. Wirel. Commun. Netw. **2019** (2019). https://doi.org/10.1186/s13638-019-1392-6
73. E. Zunic, K. Korjenic, K. Hodzic, D. Donko, Application of facebook's prophet algorithm for successful sales forecasting based on real-world data. Int. J .Comput. Sci. Inf. Technol. **12**, 23–36 (2020). https://doi.org/10.5121/ijcsit.2020.12203
74. S.A. Yarushev, A.N. Averkin, Review of studies on time series forecasting based on hybrid methods, neural networks and multiple regression. Int. J . Soft Syst. **31**, 75–82 (2016). https://doi.org/10.15827/0236-235x.113.075-082

Chapter 9
Application of Deep Learning in Recurrence Plots for Multivariate Nonlinear Time Series Forecasting

Sun Arthur A. Ojeda, Elmer C. Peramo, and Geoffrey A. Solano

Abstract We present a framework for multivariate nonlinear time series forecasting that utilizes phase space image representations and deep learning. Recurrence plots (RP) are a phase space visualization tool used for the analysis of dynamical systems. This approach takes advantage of recurrence plots that are used as input image representations for a class of deep learning algorithms called convolutional neural networks. We show that by leveraging recurrence plots with optimal embedding parameters, appropriate representations of underlying dynamics are obtained by the proposed autoregressive deep learning model to produce forecasts.

9.1 Introduction

Forecasting time series has been instrumental in providing scientific and operational insights on various problem domains. Statistical and machine learning (ML) methods are commonly used in time series modelling. Such methods include autoregressive (AR) models, vector autoregression (VAR), exponential smoothing, neural networks, and gradient boosting algorithms. Traditional statistical methods rely on significant effort in model tuning and adequate domain knowledge to obtain meaningful results. Machine learning algorithms are more flexible in terms of these modelling factors, but still require appropriate data processing treatments, which must be selected carefully with consideration of the nature of data and the problem's complexity. With the abundance of data and computational resources, deep learning (DL), a subset of machine learning, offers the most flexibility due to its capacity to learn complex

S. A. A. Ojeda · G. A. Solano (✉)
University of the Philippines Manila, Manila, Philippines
e-mail: gasolano@up.edu.ph

S. A. A. Ojeda
e-mail: saojeda@up.edu.ph

E. C. Peramo
DOST-Advanced Science and Technology Institute, Quezon City, Philippines
e-mail: elmer@asti.dost.gov.ph

G. A. Tsihrintzis et al. (eds.), *Advances in Machine Learning/Deep Learning-based Technologies*, Learning and Analytics in Intelligent Systems 23,
https://doi.org/10.1007/978-3-030-76794-5_9

feature abstractions without feature engineering, a procedure that is necessary for machine learning counterparts.

Forecasting time series can be framed as a regression task in which a forecast $\hat{\boldsymbol{x}}_{t+1}$ is modelled as $\hat{\boldsymbol{x}}_{t+1} = f(\boldsymbol{x}_t, \boldsymbol{x}_{t-1}, \boldsymbol{x}_{t-2}, \ldots, \boldsymbol{x}_{t-k-1}; \Theta)$ where f is an arbitrary model with a parameter tensor Θ learned and estimated from data and $\boldsymbol{x}_t \in \mathbb{R}^d$ is a vector containing observations at time t with dimensionality d, i.e. the number of variables. The model obtains $\hat{\boldsymbol{x}}_{t-2}$ by feeding $\hat{\boldsymbol{x}}_{t+1}$ as a new input to f taking k lagged inputs. This is repeated to obtain forecast values for the succeeding time steps. This process is known as a repeated *one-step ahead forecast*.

In this work, we develop a framework based on deep learning that builds on the one-step-ahead forecasting approach to modelling and takes advantage of time series image representations to learn discriminative features in an unsupervised manner, which allows for flexibility and convenience since it typically requires minimal data processing and analysis in order to achieve substantial results. First, evidence for non-linearity in data is obtained through statistical tests. Second, the theoretical aspects of the proposed imaging transformation are discussed. Third, the proposed model based on a convolutional neural network is outlined. Afterwards, a model training and inference pipeline is presented. Finally, we perform statistical comparisons between different configurations of the proposed model and state-of-the-art models adapted to multivariate time series regression.

9.2 Related Work

9.2.1 Background on Recurrence Plots

Time series data generated from real-world processes are generally a product of dynamical systems, which exhibits nonlinear behavior. They are usually non-stationary. These dynamical systems are an active field of study, and Eckmann et al. in [1] proposed a graphical tool called *recurrence plots* (RP) for measuring its time constancy. State-space reconstruction lays the foundation of nonlinear time series analysis. Through recurrences in the state space, this tool helps researchers analyze dynamical systems by showing information that are not easily discernible by other methods. Since then, there has been a growing interest in the topic because of its wide applicability, as most techniques would require that a time series be *stationary* or can be transformed to one by a set of arbitrary algebraic operations. This is evident in the comprehensive work of Casdagli [2] where he utilized RPs to provide a more computationally efficient and detailed characterization, prediction, change point detection, and hypothesis testing of a time series realized by dynamical systems. Another technical work about the topic was explored by Thiel et al. [3] where they expound on the amount of information contained in a recurrence plot by shedding light on the conditions for the reconstructability of the underlying dynamics from which the time series was realized, which effectively validates Takens' Theorem.

The interest in recurrence plots as a visualization tool prompted researchers to explore its interpretability and effectiveness through systematic and structural measures. This led to the development of the field called *recurrence quantification analysis* (RQA), which established useful quantitative characterizations on the structures that appear on the RP. To produce an ideal recurrence plot from a time series realization, two important parameters must be chosen. These are the amount of time delay (lag) τ and the embedding dimension m. Most of the work done in the estimation of these parameters employs meta-heuristic approaches: For the lag parameter, the most popular technique is choosing the first minimum of the logarithm of the generalized correlation integral [4] or based on the mutual information [5] between the original time series and the lagged time series. Estimating the embedding dimension m is relatively easier than estimating the lag value, and the most popular approach is the use of false nearest neighbor algorithm [6]. In 1998, Iwanski and Bradley [7] argued that the reconstructability of the hidden dynamics of a time series appear to be independent of the embedding dimension as derived from several experimental data. Their experiments suggest that qualitative features that are visible in an RP using a particular value of m persists or are also visible in another RP with a different embedding dimension. Gao and Cai [8] did further studies on RPs exploring their patterns and structures on the different values of the embedding parameters. They were able to identify patterns and derived the features from the time series that generated those patterns. Based on differential entropy, Gautama et al. [9] hypothesized that optimal values for both the lag and the embedding dimension can be simultaneously determined. With the introduction of random surrogate data, they proposed an entropy ratio method to compute these values. Application of RPs and RQA goes beyond univariate time series. This can be seen in a paper by Marwan et al. [10] where they proposed a generalized method in higher-dimensional spatial data.

Aside from the lag value and the embedding dimension, another parameter in RP that controls its overall qualitative characteristics is the *threshold* ε. Sipers [11] et al. studied unthresholded recurrence plots and determined theoretically and analytically which information can and cannot be extracted from a signal. They validated their results experimentally from EEG data. Penatti et al. [12] used RPs for human activity recognition based on mobile inertial sensors. They computed RPs from sensor data and applied machine learning techniques for prediction. More and application-centric of RP research have been published in recent years. Noteworthy is the work of Marwan [13], who compiled a survey of the different research and practical applications and review how this interesting field had advanced and proliferated over the years. Since its invention more than two decades ago, more than a hundred research papers on recurrence plots have been published in scientific and engineering domains that require extensive analyses of nonlinear dynamics.

9.2.2 *Time Series Imaging and Convolutional Neural Networks*

Wang and Oates [23] proposed two algorithms for time series imaging: *Gramian Angular Fields Fields* (GAF) and *Markov Transition Fields* (MTF). Both of these algorithms take time series data to produce their respective image representations. Gramian Angular Field takes advantage of polar coordinates to encode radial and angular information in the Gramian Matrix, which is essentially an image matrix. Markov Transition Field, on the other hand, identifies quantile bins and then encodes their pairwise transition probabilities in a markov transition matrix representing an image. These images are fed into a *Tiled Convolutional Neural Network* (Tiled CNN) that is used in classification problems. Benchmarks reveal that the GAF-MTF approach outperforms 1NN, SAX-VSM, and other state-of-the-art machine learning algorithms in 12 datasets from http://timeseriesclassification.com.

Hatami et al. [24] demonstrated the use of recurrence plots (RP) and convolutional neural networks for time series classification. They proposed a pipeline to convert time series into texture images using recurrence plots that are fed to the model to produce predictions. Results show that in 20 univariate time series classification datasets from the UCR archive, the method outperforms approaches that rely on time series encoding such as GAF and MTF.

Several works focused on applying recurrence plots and convolutional neural networks were motivated by time series imaging using recurrence plots after having achieved state-of-the-art results in various domains. Hsueh [25] applied recurrence plot imaging and used a convolutional neural network model to diagnose induction motor's fault conditions including bearing axis deviation and stator and rotor friction for monitoring induction motor health and obtained 99.81% validation accuracy. Using a similar approach for fetal hypoxia diagnosis, Zhao et al. [26] implemented a Computer-Aided Diagnosis System that also utilizes a convolutional neural network to obtain a 10-fold cross-validation accuracy of 98.69%. Finally, for human activity recognition based on raw sensor data, Ceja et al. [27] achieved better results using recurrence plots against models that use feature-extraction methods such Deep Belief Network and Multi-layer Perceptron by obtaining a 10-fold cross-validation accuracy of 94.2%. Finally, we extend the methodology of our previous work [28] designed for time series classification applied to the meteorological domain. In this work, our framework for multivariate nonlinear time series forecasting can be used in any domain, as long as the forecasting task can be appropriately represented as described in the succeeding sections.

9.3 Time Series Nonlinearity

Data complexity is an important factor that fundamentally affects time series modelling. However, it can mean differently across studies as it is defined depending on the objective of the researchers. For instance, in machine learning, data complexity is defined such that it is related to and is dependent on the classification algorithm. These sort of data complexity measures can be used in order to identify which model has the greatest performance among other classification models that is being compared to. It can also be used to understand how capable a classification algorithm is in modelling its input data. Several data complexity measures have been reviewed by Sotoca et al. in [15] particularly designed for classification models. Li and Abu-Mustafa [16] developed several measures based on Kolmogorov complexity. These methods, however, are not model-agnostic. In this research, data complexity must be defined such that it describes a particular dataset without involving the model. To characterize the inherent complexity in time series data that is independent of the model in question, we consider *time series nonlinearity* as an ideal specification of data complexity. Nonlinearity in time series is of particular interest to researchers since most natural and artificial processes in chaos-theoretic domains exhibit this characteristic. Nonlinear dynamics generally entails that a process has considerable complexity that may be difficult to analyze and model. Moreover, these problems are also often multivariable, wherein such variables (parameters) directly influence the behavior of a system being considered.

A univariate time series $\{x_t\}$ is *linear* if it that can be represented as

$$x_t = \sum_{j=-\infty}^{\infty} \theta_j \varepsilon_{t-j}$$

where $\{\varepsilon_t\}$ are independent and identical distribution (i.i.d.) random variables with finite variance. A nonlinear time series is one that does not satisfy this representation. To identify if there exists nonlinearity in a multivariate time series $\{\boldsymbol{x}_t\}$ with d variables, each univariate time series $\{\boldsymbol{x}_{it}\}$, $i = 1, \ldots, d$ is tested for linearity.

The Brock, Dechert, and Scheinkman (BDS) test [17–19] is designed to test the null hypothesis of i.i.d. To test for linearity, the BDS test is applied to the residuals of a linear model fitted on $\{\boldsymbol{x}_{it}\}$. If the null hypothesis is rejected, the model is misspecified. This departure can be interpreted as the detected presence of nonlinearity unaccounted for by the model.

Consider $\{x_t\}$, the correlation integral is estimated by

$$c_{m,n}(\epsilon) = \frac{2}{(n-m+1)(n-m)} \sum_{t=s+1}^{n} \sum_{s=m}^{n} \prod_{j=0}^{m-1} I_\epsilon(x_{t-j}, x_{s-j})$$

where n is the sample size, m is the embedding dimension, and I_ϵ is the indicator function

$$I_\epsilon = \begin{cases} 1 & \text{if } |x_{t-j} - x_{s-j}| < \epsilon \\ 0 & \text{otherwise} \end{cases}$$

which is used to estimate the joint probability of two m-dimensional points within a distance ϵ

$$\Pr(|x_t - x_s| < \epsilon, |x_{t-1} - x_{s-1}| < \epsilon, \ldots, |x_{t-m+1} - x_{s-m+1}| < \epsilon),$$

note that under the null hypothesis, $\mathrm{E}(c_{m,n}(\epsilon)) = (\mathrm{E}(c_{1,n}(\epsilon)))^m$.

The BDS test statistic is defined as

$$w_{m,n}(\epsilon) = \sqrt{n-m+1}\frac{c_{m,n}(\epsilon) - c_{1,n-m+1}^m(\epsilon)}{\sigma_{m,n}(\epsilon)}$$

where $\sigma_{m,n}(\epsilon)$ is the standard deviation of $\sqrt{n-m+1}(c_{m,n}(\epsilon) - c_{1,n-m+1}^m(\epsilon))$ that can be estimated as detailed in [17].

In the BDS test, we use the residuals $\{e_t\}$ of an Autoregressive Integrated Moving Average (ARIMA) model to test for i.i.d. The ARIMA(p, d, q) model is defined as

$$x'_t = \mu + \sum_{i=1}^{p} \alpha_i x'_{t-i} + \sum_{j=1}^{q} \beta_j x'_{t-j} + \varepsilon_t$$

where $\{x'_t\}$ is the d-th differenced series, p is the autoregressive (AR) order, and q is the moving average (MA) order [20]. To eliminate manual procedures in finding the optimal p, d, q, the Hyndman-Khandakar algorithm can be used. Hyndman-Khandakar algorithm is a stepwise search algorithm that combines AICc minimization and repeated KPSS tests to obtain the optimal p, d, q automatically. The Hyndman-Khandakar algorithm [29] is outlined as follows:

1. Try the following four possible models.
   - ARIMA$(2, d, 2)$ if $m = 1$ and ARIMA$(2, d, 2)(1, D, 1)$ if $m > 1$
   - ARIMA$(0, d, 0)$ if $m = 1$ and ARIMA$(0, d, 0)(0, D, 0)$ if $m > 1$
   - ARIMA$(1, d, 0)$ if $m = 1$ and ARIMA$(1, d, 0)(1, D, 0)$ if $m > 1$
   - ARIMA$(0, d, 1)$ if $m = 1$ and ARIMA$(0, d, 1)(0, D, 1)$ if $m > 1$

   If $d + D \leq 1$, fit models with $c \neq 0$. Otherwise, set $c = 0$. Select the one with the smallest AIC value. This is called the "current" model and is denoted by ARIMA(p, d, q) if $m = 1$ or ARIMA$(p, d, q)(P, D, Q)_m$ if $m > 1$.
2. Consider the following current model variations:
   - where one of p, q is allowed to vary by ±1 from the current model;
   - where p and q both vary by ±1 from the current model;
   - where P and Q both vary by ±1 from the current model;
   - where the constant c is included if the current model has $c = 0$ or excluded if the current model has $c \neq 0$.

   When model with lower AIC is found, consider it as the current model and repeat the procedure. The algorithm terminates when a model close to the current model with lower AIC cannot be found.

It should be noted that this algorithm implements the case of seasonal ARIMA (SARIMA). For simplicity and convenience in determining nonlinearities, it is not advised that the SARIMA be used in this approach. The underlying principle in determining nonlinearity in this approach is that it must be simple and automatic so that it will not defeat its purpose of minimizing human intervention. Performing extraneous procedures to identify whether or not the data is influenced by seasonality can be a time-consuming process and will require considerable domain knowledge on the dataset being considered.

9.4 Time Series Imaging

Imaging time series by encoding the raw time series into image representations is a useful tool in the analysis of temporal patterns and dominant characteristics. For instance, one may particularly be interested in characterizing these features in order to identify properties that might explain the behavior of a nonlinear deterministic system with respect to time. In seismological, neural, speech processing and related fields, spectrograms are used to analyze, identify and characterize the relationship between measurements and events in a time series using a frequency-domain representation. Similarly, a *recurrence plot* (RP) is a 2-dimensional visualization tool introduced by Eckmann et al. [1] for dynamical systems and non-stationary time series that highlights the state recurrence in a multi-dimensional embedding called the phase space trajectory. It is primarily used for visual inspection of a system's topological and textural properties and its characterization using *recurrence quantification analysis* (RQA). Figure 9.1 illustrates example recurrence plot representations generated from three arbitrary time series data.

The recurrence plot $\boldsymbol{R}$ of a univariate time series $\{x_t\}$ is constructed from the delay vectors $\boldsymbol{x}_i$

$$\boldsymbol{x}_i = (x_i, x_{i+\tau}, \ldots, x_{i+(m-1)\tau}),\ i = 1, \ldots, n-(m-1)\tau$$
$$\boldsymbol{R}_{i,j} = H(\varepsilon - \|\boldsymbol{x}_i - \boldsymbol{x}_j\|), i,\ j = 1, \ldots, n-(m-1)\tau$$

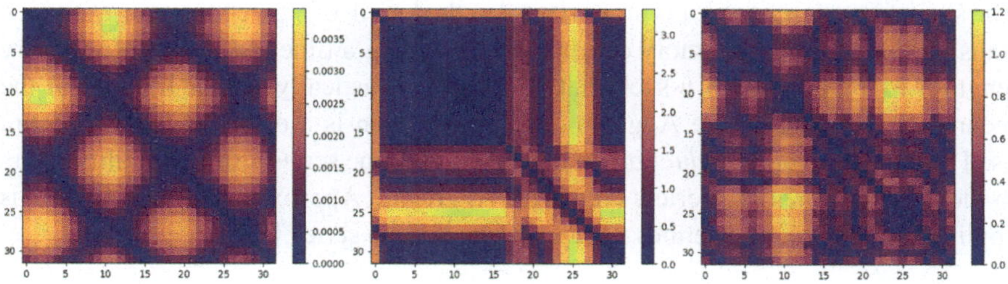

Fig. 9.1 Recurrence plots generated from time series data

where m is the embedding dimension, τ is the time delay, n is the length of the time series, ε is the threshold value, and H is the heaviside function.

Time series imaging must be performed on time series data to produce representations that are suitable as input to a convolutional neural network. Recurrence plots can be used to generate such image representations. To motivate the use of recurrence plots over imaging techniques based on spectral analysis, any strong limitation in the alternative method (i.e. spectrogram) must be discussed.

Spectrograms assume signal stationarity. Intuitively, a *stationary* time series is a time series wherein its statistical characteristics do not vary over time (time-invariance). A less strict formal definition of stationarity (weak stationarity) is a time series $\{x_t\}$ that satisfies constant mean in all time steps $\mu = \mu_0,\ t = 1, \ldots, T$ and there exists $f : \mathbb{Z} \to \mathbb{R}$ such that $f(\ell) = \mathrm{cov}(x_t, x_{t+\ell}), \forall t, \ell$. In other words, the covariance between x_t and $x_{t+\ell}$ depends only on lag difference ℓ for all t and ℓ [21]. Time-dependent variance and asymmetric cycles exist in nonlinear time series which violates the stationarity assumption. In many real-world time series, stationarity is not met, especially when the process exhibits nonlinearity. Pesaran [22] suggested that weakening the stationarity assumption to short-time stationarity is a reasonable procedure. Short-time stationarity assumes that stationarity exists within short time frames, and changes in longer time scales. This inevitably leads to a modelling implementation that requires manual analysis of the time series considered.

The addition of a manual procedure is generally undesirable since it unnecessarily complicates the process by requiring supervision which defeats the purpose of the proposed approach. Thus, the use of spectrograms is inappropriate for the purpose of time series modelling since it will, as a consequence of violating the stationarity assumption, be prone to information loss. On the other hand, the use of recurrence plots will allow the convolutional neural network to extract discriminative features based on its local and global spatial properties and structure arising from the representation, since it is designed to encode temporal information from a nonlinear time series.

9.4.1 Dimensionality Reduction

The size of a recurrence plot is proportional to the length of a given time series, which translates to an $\mathcal{O}(n^2)$ memory complexity. Reducing a time series $\{x_t\}$ of arbitrary length n to a fixed length $\bar{n}$ such that $\bar{n} \ll n$ will drastically reduce the size of the recurrence plot. Piecewise Aggregate Approximation is a deterministic algorithm used for *dimensionality reduction*. In the context of time series, dimensionality refers to the length of the time series. Piecewise Aggregate Approximation (PAA) [31] is applied to $\{x_t\}$ which generates the approximate time series $\{\overline{x}_t\}$ by

$$\bar{x}_i = \frac{\bar{n}}{n} \sum_{j=n/\bar{n}(i-1)+1}^{(n/\bar{n})i} x_j$$

The recurrence plot $\boldsymbol{R}$ is instead constructed from $\{\bar{x}_t\}$. To this end, the chosen $\bar{n}$ for the Piecewise Aggregate Approximation algorithm can be thought of as a hyperparameter that controls the image size (dimensions) of a recurrence plot $\boldsymbol{R} \in \mathbb{R}^{\bar{n}\times\bar{n}}$. It is clear that as $\bar{n} \to n$, the greater amount of information is encoded to $\boldsymbol{R}$, which may improve the model's performance. Thus, $\bar{n}$ must be chosen carefully according to the hardware (memory) budget available and the problem's inherent complexity. If the nature of $\{x_t\}$ can be characterized as one that contains finer subtleties (local features), then $\bar{n}$ must be relatively large to effectively encode these information. Similarly, if $\{x_t\}$ contains larger and more global features, $\bar{n}$ can be smaller, as these features can be sufficiently captured in the approximation $\{\overline{x}_t\}$.

9.4.2 *Optimal Parameters*

In order to obtain a consistent image representation of a time series $\{x_t\}$, optimal recurrence plot parameters threshold value ε, embedding dimension m, and time delay τ specific to $\{x_t\}$ must be set to ensure that its respective recurrence plot reflects its dynamics appropriately. The threshold value ε controls the amount of recurrences and recurrent structures that occurs in the plot. For all variables, we set this to $\varepsilon = 0$, since determining the threshold value is a non-trivial task that heavily depends on the objective of the researcher [32]. To the best of our knowledge, literature proposing optimal threshold values only provides recommendations for (manual) recurrence analysis, not when recurrence plots are used in an automated modelling approach that utilizes deep learning. Moreover, appropriate embedding dimension m and time delay τ must be set so as to avoid interruptions and artificial increase in diagonal lines and small blocks [32]. Non-optimal m and τ values also result in phase space (delay) vectors that are dominated by noise [9].

For time series from discrete-time maps, Cao recommends a time delay $\tau = 1$ [30]. Since the approximate time series $\{x_t\}$ generated by the Piecewise Aggregate Approximation algorithm can be interpreted as the mapping $\mathbb{R}^n \to \mathbb{R}^{\bar{n}}$, $\{x_t\}$ approximated as $\{\bar{x}_t\}$ qualifies as an output of a discrete-time map. Notice that setting $\tau = 1$ avoids large skips between elements x_i and $x_{i+\tau}$ in the delay vector $\boldsymbol{x}_i$ as these skips in x_t affect the granularity of information in $\boldsymbol{x}_i$. Relatively large skips as a consequence of setting $\tau > 1$ over $\{\bar{x}_t\}$ in generating $\boldsymbol{R}$ can be problematic, since each element in $\{\bar{x}_t\}$ is a representative of a particular segment in $\{x_t\}$. In other words, setting $\tau = \tau_0 > 1$ translates to skipping $\tau_0 - 1$ points for every $\bar{x}_i$ in $\{\bar{x}_t\}$, which generally leads to information loss.

Finally, following the findings of Iwanski and Bradley [7] that recurrence plots of experimental data appear to be independent of the embedding dimension m, it is set to $m = 1$. Iwanski and Bradley demonstrated that increasing m does not affect the recurrence plot both qualitatively and quantitatively. This result is important since deep learning relies on pattern recognition. If topological structures remain consistent or very similar across varied m, then the process of identifying an appropriate value for m using various heuristics can be avoided. This significantly simplifies the overall methodology and modelling process since the optimal embedding dimension and time delay values ($m = 1, \tau = 1$) are already identified using both theoretical and experimental bases without the need for any further estimation.

9.5 Convolutional Neural Networks

Motivated by the need to adapt traditional neural networks to image classification and regression tasks, convolutional neural networks were developed. Consider a classification task wherein the input is an $n \times n$ image flattened to form a vector of length n^2. Let F be a feedforward neural network with $n_0 = n^2$ input units. For each layer in a forward pass, the matrix multiplication $\boldsymbol{\Theta}^\top \boldsymbol{x}$ of layer's weights $\boldsymbol{\Theta} \in \mathbb{R}^{n \times m}$ and input from the previous layer $\boldsymbol{x} \in \mathbb{R}^{n \times 1}$ is an $\mathcal{O}(mn)$ operation. When n_0 is very large, the matrix multiplication in the immediate hidden layer becomes an issue in terms of both memory and time complexity. In addition, suppose that a binary image $\boldsymbol{I}$ represented by a matrix contains a local structure represented by a group of 1's centered at (x_0, y_0) and an image $\boldsymbol{I}'$ containing the same local structure translated to a position centered at $(x_0 - c, y_0 - c)$. F will treat I and I' as images containing their own unique local structure. Thus, computing for $\boldsymbol{y} = h(\boldsymbol{\Theta}^\top \boldsymbol{x} + \boldsymbol{b})$ using I versus computing $\boldsymbol{y}$ using I' will yield different outputs. As a consequence, F would not efficiently recognize similar and redundant patterns across instances in the data resulting to the need to increase the model's complexity by means of increasing the number of hidden units and layer depth. These issues can be resolved with the use of a convolutional neural network due to its properties that are useful for learning spatial data: *sparse interactions*, *weight sharing*, and *equivariance*. These are made possible by introducing *convolutional* and *pooling* layers.

A convolutional layer is based on the discrete *convolution* operation. A discrete convolution denoted by $\circledast$ is defined as

$$[f \circledast g](i) = \sum_a f(a)g(i-a)$$

For 2-dimensional inputs such as images, we define the discrete convolution over the two axes of an image $\boldsymbol{I}$ [33, 34]

$$[\boldsymbol{I} \circledast \boldsymbol{K}](i, j) = \sum_p \sum_q \boldsymbol{I}(i, j)\boldsymbol{K}(u-p, v-q)$$

where $\boldsymbol{K} \in \mathbb{R}^{k \times k}$ is called the *kernel*, a matrix of learnable weights. Convolutions are commutative, so the above equation is equivalent to

$$[\boldsymbol{I} \circledast \boldsymbol{K}](i, j) = \sum_p \sum_q \boldsymbol{I}(i-p, j-q)\boldsymbol{K}(u, v)$$

By making kernel $\boldsymbol{K}$ smaller than $\boldsymbol{I}$, sparse connectivity is achieved, which reduces the time complexity to $\mathcal{O}(kn)$ versus $\mathcal{O}(mn)$, when $k \ll m$. This is also an improvement in memory complexity, since it drastically reduces the number of parameters that are needed to be stored. Another advantage of using convolutions is parameter sharing, which allows for the use of the same parameters for different inputs. As a consequence, parameter sharing in a convolution causes the convolutional layer to be *equivariant*. Mathematically, a function f is equivariant to a function g if $f(g(x)) = g(f(x))$. This makes the convolutional neural network tolerant to translations in $\boldsymbol{I}$, in contrast to a feedforward neural network that do not reuse weights, which also is a major advantage in memory complexity.

Furthermore, as the depth of a convolutional neural network increases, it is desirable to reduce the sizes of outputs per layer in order to reduce the computational and statistical burden of succeeding layers. The reduction is a downsampling operation accomplished using a pooling layer. A pooling operation essentially takes the maximum (max pooling) or the average (average pooling) of a submatrix $\boldsymbol{x}_{\subseteq} \in \mathbb{R}^{s \times s}$ of an input matrix $\boldsymbol{x} \in \mathbb{R}^{H \times W}$ and placing it as a cell in an output matrix $\boldsymbol{z} \in \mathbb{R}^{H' \times W'}$ collecting maximum or average values per submatrix. Let the submatrix positioned topmost left at (w, h) be denoted by $\boldsymbol{x}_{w:w+s,h:h+s}$. Maximum and average pooling are defined as the following, respectively

$$z_{i',j'} = \max_{0 \le i,j < s} \{\boldsymbol{x}_{i':H+i,j':W+j}\}$$

$$z_{i',j'} = \frac{1}{s^2} \sum_{0 \le i,j < s} \{\boldsymbol{x}_{i'+i,j'+j}\}$$

In the same way that the layers in a feedforward neural network are stacked, a convolutional neural network is stacked from left to right, incorporating the convolutional and pooling layers

$$\boldsymbol{x} \rightarrow \boxed{\circledast} \xrightarrow[\phi]{} \boxed{\odot} \rightarrow \ldots \rightarrow \boxed{\circledast} \xrightarrow[\phi]{} \boxed{\odot} \xrightarrow[\text{flat}]{} \boxed{\oplus} \xrightarrow[\phi]{} \ldots \xrightarrow[\phi]{} \boxed{\oplus} \xrightarrow[\text{lin}]{} \boldsymbol{y}$$

In the figure above, input $\boldsymbol{x}$ is transformed by a series of convolutional ($\circledast$) and pooling ($\odot$) layers mediated by a nonlinear activation function ϕ, flattened (denoted by *flat*), and connected to a fully-connected layer ($\oplus$) where the final layer outputs $\boldsymbol{y}$ with a linear activation function (denoted by *lin*). For brevity, $\boldsymbol{x}$ is a *single-channel* input. For multivariate inputs, $\boldsymbol{x} \in \mathbb{R}^{H \times W}$ is replaced with an tensor $\boldsymbol{X} \in \mathbb{R}^{H \times W \times D}$, where the last axis represents the number of variables D (channels). The process essentially remains the same despite the inclusion of the channel axis to handle the multivariate case.

9.6 Model Pipeline and Architecture

For convenience, we describe the pipeline using a training instance from the dataset $\mathcal{D} = \{(\boldsymbol{X}_{t-k:t}, \boldsymbol{x}_{t+1})\}$. The procedure begins with the transformation of a multivariate time series window $\{\boldsymbol{X}_i\}_{t-k:t}$ into a tensor $\boldsymbol{X}_{t-k:t} \in \mathbb{R}^{R\times R\times D}$ containing recurrence plots of each variable where R is the image size and D is the number of variables. This is done by applying Piecewise Aggregate Approximation and time series imaging using recurrence plots to each univariate time series $\{\boldsymbol{X}_{it}\}_{t-k:t}$ representing variable i. Next, the model takes $\boldsymbol{X}_{t-k:t}$ as input and is forward-propagated to obtain the output $\hat{\boldsymbol{x}}_{t+1}$, which is the model's prediction of the label vector containing one-step ahead values for each variable $\boldsymbol{x}_{t+1} = \{\boldsymbol{X}_t\}_{t:t+1} \in \mathbb{R}^{D\times 1}$.

9.6.1 *Architecture*

The proposed model architecture is outlined in Table 9.1. The model consists of two convolutional and fully-connected (FC) layers. *Batch normalization* [35] is used to accelerate model training and acts as an additional regularization component by normalizing layer inputs. Dropouts [36] are also included to reduce the risk of overfitting by randomly dropping nodes along with their connections during training.

9.7 Experimental Setup

The model is compared against the following state-of-the-art models: ResNet [37], Fully Convolutional Networks [38], and InceptionTime [39]. We trained three models, CNN-16, CNN-32, CNN-64, each corresponds to the same model architecture

Table 9.1 Proposed model architecture

Type	Size/rate	# Filters/nodes	Activation
Conv + Batch norm	3×3	32	Relu
Max pooling	2 × 2	–	–
Conv + Batch norm	3 × 3	32	Relu
Max pooling	2 × 2	–	–
Dropout	0.25	–	–
Flatten	–	–	–
FC	–	256	Relu
Dropout	0.25	–	–
FC	–	D	Lfinear

Table 9.2 Selected datasets from http://timeseriesregression.org

Dataset	No. of variables with $p < 0.05$
BIDMC32HR	2
BenzeneConcentration	5
HouseholdPowerConsumption1	5
IEEEPPG	2
LiveFuelMoistureContent	7
NewsHeadlineSentiment	2
PPGDalia	4

trained with different image size configurations [16, 32]. The following models are used to test if there is an improvement in performance if the same generic model architecture is used. We used the following datasets from http://timeseriesregression.org [40] (Table 9.2):

The criteria for dataset selection are based on the following:

- Number of variables must be greater than 1.
- For convenience, time series length must at least be greater than 64. This allows the use of the CNN-64 model without any extrapolation procedures on datasets with a length less than 64.
- Training size must at least contain 1000 instances.
- Datasets containing essentially the same training samples are discarded. For instance, BIDMC32HR, BIDMC32RR and BIDMC32SpO2 only differs in their labels. Hence, only BIDMC32HR is retained.
- At least one variable exhibits nonlinearity.

Minimal data preprocessing has been performed in the datasets. In the case of missing values, linear interpolation was performed. The BDS test (discussed in Sect. 3) was applied to each dataset to determine the number of variables that exhibit nonlinearity. Each model is trained for 1000 epochs with a batch size of 1024. *Early stopping* and *learning rate reduction* on plateau is performed to prevent overfitting. Once the model is trained, it is validated against the testing set with Root Mean Squared Error (RMSE) as the metric averaged over D variables

$$\text{RMSE} = D^{-1}\sqrt{n^{-1}\sum(\hat{\boldsymbol{x}}_{t+1} - \boldsymbol{x}_{t+1})^2}$$

To ensure consistency, for each dataset we trained each model for 5 runs and averaged the results. To compare the models, we performed Friedman's test [41], which is used to detect any differences in average ranks between models. In case of ties, fractional ranking is used. Given k models and b different datasets, it tests for the null hypothesis $H_0 : M_1 = M_2 = \ldots = M_k$ and the alternative hypothesis $H_a : M_i' \neq M_i$ for at least one $i' \neq i$ where M_i's are the median responses associated with the k models. Let $R_i j$ be the rank of the i-th model within the j-th dataset and let R_i be the sum of the ranks for the i-th model for all datasets. The Friedman test statistic is computed by

$$\chi_F^2 = \frac{12}{bk(k+1)} \sum_{i=1}^{k} \left[R_i - \frac{b(k+1)}{2} \right]^2$$

where $\frac{b(k+1)}{2} = E_0(R_i)$ is the mean of the R_i's when H_0 is true. If H_0 is false, then a post-hoc test called the Nemenyi test is conducted to identify which model(s) perform significantly different among others. The Nemenyi test [42] computes for the critical difference

$$CD = q_\alpha \sqrt{\frac{k(k+1)}{6N}}$$

Any model that differs by at least CD with $\alpha = 0.05$ outperforms other models based on the average ranks shown in a critical difference diagram.

`Tensorflow` was used to implement the models. ResNet, FCN, and InceptionTime implementations are adapted from [40]. We also used the `pyts` package to implement recurrence plots and the Piecewise Aggregate Approximation algorithm. The BDS test is implemented using the `statsmodels` package and the automatic ARIMA hyperparameter optimization using the `pmdarima` package. All of these were implemented in Python 3.8. The implementation is available at https://github.com/saojeda/rcnn-experiments.

9.8 Results

Table 9.3 shows the results of the experiments. Based on this table, we computed the Friedman test statistic and corresponding p-value. Friedman's test show that H_0 cannot be rejected at $p = 0.69220 > 0.05$, hence the post-hoc Nemenyi test used to visualize the critical difference diagram was not conducted.

The models CNN-16, CNN-32, CNN-64 do not outperform FCN, InceptionTime, and ResNet. However, the CNN models' performance are still on par with these state-of-the-art time series models in one-step ahead forecasting tasks. It is worth noting that the CNN models appear to perform better in datasets resembling discrete-time maps. It is evident that the FCN, InceptionTime, and ResNet models perform better compared with the CNN models in the datasets BIDMC32HR, IEEEPPG, and PPGDalia which are sampled in an interval resembling continuous-time. It is reasonable to infer that these models may have local structures that are not sufficiently captured by the Piecewise Aggregate Approximation algorithm given length $\bar{n}$. The CNN models may perform better with a larger $\bar{n}$, but available memory resource

Table 9.3 Model training results showing the average RMSE over 5 runs of a model in a particular dataset. Long dataset names are shortened

	CNN-16	CNN-32	CNN-64	FCN	InceptionTime	ResNet
BIDMC32HR	0.35622	0.34296	0.36488	0.10863	0.13123	**0.10863**
BConcentration	292.79553	258.91063	**241.59186**	600.77448	659.98979	780.95314
HPConsumption1	4.66115	**4.08298**	4.12137	101.34325	95.23767	103.55190
IEEEPPG	42.15604	43.63521	41.84182	63.89978	**26.56151**	45.42202
LFMContent	1575.85710	1414.00837	**1223.33757**	1835.93391	1927.07244	1835.93391
NHSentiment	**14.98553**	15.79095	16.04000	15.01115	14.98857	14.99808
PPGDalia	55.00934	51.68706	43.66140	**1.61548**	1.62459	**1.61548**

can be a constraint. In these cases, it is desirable to use the alternative State-of-The-Art (SoTA) models. The CNN models were able to provide significant boosts in performance particularly in datasets that are by nature sampled in discrete-time, with several orders of magnitude of improvement versus SoTA time series models.

9.9 Conclusion

In this work, we presented a framework that makes use of recurrence plots in order to perform multivariate nonlinear one-step ahead forecasts that can be applied in real-world use cases. Raw time series data are first transformed into recurrence plot representations, wherein these image representations are learned by a generic convolutional neural network model to recognize discriminative features based on the topological structure of the input recurrence plots. We have shown that the convolutional neural network is able to learn these patterns and consequently is able to produce forecasts in an automated manner which does not require manual data analyses and considerable data preprocessing. In addition, the three identical models varying only in image size configuration, we demonstrated that the approach is effective when it is used in time series data that are discrete in nature. This is due to the consequence of dimensionality reduction that is beneficial when applied to shorten time series data with fewer local structures and greater global structures. This observation can help researchers determine when to apply this approach. In general, the authors recommend that this approach is applied to problems with discrete-time nonlinear variables. Time series that are continuous in nature can be modelled better with alternative SoTA methods in which the proposed approach was compared against. In the future, the authors could work on extending this approach to more specialized convolutional neural network architectures that can take advantage of the topological properties of recurrence plots to further improve its performance.

References

1. J.P. Eckmann, S. Oliffson Kamphorst, D. Ruelle, Recurrence plots of dynamical systems. Europhys. Lett. **4**(9), 973–977 (1987). https://doi.org/10.1209/0295-5075/4/9/004
2. M.C. Casdagli, Recurrence plots revisited. Phys. D **108**(1–2), 12–44 (1997). https://doi.org/10.1016/S0167-2789(97)82003-9
3. M. Thiel, M.C. Romano, J. Kurths, How much information is contained in a recurrence plot? Phys. Lett. A **330**(2004), 343–349 (2004). https://doi.org/10.1016/j.physleta.2004.07.050
4. W. Liebert, H.G. Schuster, Proper choice of the time delay for the analysis of chaotic time series. Phys. Lett. A **142**(2, 3), 107–111 (1989). https://doi.org/10.1016/0375-9601(89)90169-2
5. A.M. Fraser, H.G. Swinney, Independent coordinates for strange attractors from mutual information. Phys. Rev. A **33**(2), 1134 (1986). https://doi.org/10.1103/PhysRevA.33.1134
6. M.B. Kennel, R. Brown, H.D.I. Abarbanel, Determining embedding dimension for phase-space reconstruction using a geometrical construction. Phys. Rev. A **45**, 3403 (1992). https://doi.org/10.1103/PhysRevA.45.3403
7. J.S. Iwanski, E. Bradley, Recurrence plots of experimental data: to embed or not to embed. Chaos **8**, 861–871 (1998). https://doi.org/10.1063/1.166372
8. J. Gao, H. Cai, On the structures and quantification of recurrence plots. Phys. Lett. A **270**(1–2), 75–87 (2000). https://doi.org/10.1016/S0375-9601(00)00304-2
9. T. Gautama, D.P. Mandic, M.M. Van Hulle, A differential entropy based method for determining the optimal embedding parameters of a signal, in *2003 IEEE International Conference on Acoustics, Speech, and Signal Processing, 2003, Proceedings, (ICASSP '03)* (Hong Kong, 2003), pp. VI–29. https://doi.org/10.1109/ICASSP.2003.1201610
10. N. Marwan, J. Kurths, P. Saparin, Generalised recurrence plot analysis for spatial data. Phys. Lett. A **360**(2007), 545–551 (2007). https://doi.org/10.1209/0295-5075/4/9/004
11. A. Sipers, P. Borm, R. Peeters, On the unique reconstruction of a signal from its unthresholded recurrence plot. Phys. Lett. A **375**(2011), 2309–2321 (2011). https://doi.org/10.1016/j.physleta.2011.04.040
12. O.A.B. Penatti, M.F.S. Santos, Human activity recognition from mobile inertial sensors using recurrence plots (2017). ArXiv. Available https://arxiv.org/pdf/2004.13408.pdf. Accessed 31 Aug 2020
13. N. Marwan, A historical review of recurrence plots. Eur. Phys. J. Spec. Top. **164**(2008), 3–12 (2008). https://doi.org/10.1209/0295-5075/4/9/004
14. B. Lim, S. Zohren, Time series forecasting with deep learning: a survey (2020). ArXiv. Available, https://arxiv.org/pdf/2004.13408.pdf. Accessed 24 Aug 2020
15. J.M. Sotoca, J.S. Sánchez, R.A. Mollineda, A review of data complexity measures and their applicability to pattern classification problems, in *Actas del III Taller Nacional de Mineríade Datos y Aprendizaje* (TAMIDA, 2005), pp. 77–83
16. L. Li, Y.S. Abu-Mostafa, Data complexity in machine learning. Caltech Computer Science Technical Reports (2006). https://doi.org/10.7907/Z9319SW2
17. W.A. Brock, J.A. Scheinkman, W.D. Dechert, B. LeBaron, A test for independence based on the correlation dimension. Econ. Rev. **15**(3), 197–235 (1996)
18. J. Belaire-Franch, D. Contreras, How to compute the BDS test: a software comparison. J. Appl. Econ. **17**, 691–699 (2002). https://doi.org/10.1002/jae.679
19. E. Zivot, Nonlinear time series models (n.d.). https://faculty.washington.edu/ezivot/econ584/notes/nonlinear.pdf. Accessed 24 Aug 2020
20. R.J. Hyndman, G. Athanasopoulos, ARIMA Models, in *Forecasting: Principles and Practice*, 2nd edn. (OTexts, Melbourne, Australia, 2018), pp. 221–274
21. S.S. Rao, A course in time series analysis (2020). https://www.stat.tamu.edu/~suhasini/teaching673/time_series.pdf. Accessed 29 Aug 2020
22. B. Pesaran, Spectral analysis for neural signals (2020). https://neurophysics.ucsd.edu/courses/physics_173_273/SfN_Short_Course_c1.pdf. Accessed 29 Aug 2020

23. Z. Wang, T. Oates, Visual interpretation of hand gestures for human computer interaction: a review, in *Paper presented at the Proceedings of the Twenty-Fourth International Joint Conference on Artificial Intelligence*. (Buenos Aires, Argentina, 2015). 25–31 July 2015
24. N. Hatami, Y. Gavet, J. Debayle, *Paper presented at the Tenth International Conference on Machine Visio* (Vienna, Austria, 2017). 13–15 Nov 2017
25. Y. Hsueh, V.R. Ittangihala, W. Wu et al., Condition monitor system for rotation machine by CNN with recurrence plot. Energies **12**, 3221 (2019). https://doi.org/10.3390/en12173221
26. Z. Zhao, Y. Zhang, Z. Comert, Y. Deng, Computer-aided diagnosis system of fetal hypoxia incorporating recurrence plot with convolutional neural network. Front. Physiol. (2019). https://doi.org/10.3389/fphys.2019.00255
27. E. Garcia-Ceja, M.Z. Uddin, J. Torresen, Classification of recurrence plot' distance matrices with a convolutional neural network for activity recognition. Procedia Comput. Sci. **130**, 157–163 (2018). https://doi.org/10.1016/j.procs.2018.04.025
28. S.A.A. Ojeda, G.A. Solano, E.C. Peramo, Multivariate time series imaging for short-term precipitation forecasting using convolutional neural networks, in *2020 International Conference on Artificial Intelligence in Information and Communication (ICAIIC)* (2020). https://doi.org/10.1109/icaiic48513.2020.9065238
29. R.J. Hyndman, Y. Khandakar, Automatic time series forecasting: the forecast package for R. J. Stat. Softw. (2008). https://doi.org/10.18637/jss.v027.i03
30. L. Cao, Practical method for determining the minimum embedding dimension of a scalar time series. Phys. D: Nonlinear Phenom. **110**, 43–50 (1997). https://doi.org/10.1016/s0167-2789(97)00118-8
31. E. Keogh, K. Chakrabarti, M. Pazzani, S. Mehrotra, Dimensionality reduction for fast similarity search in large time series databases. Knowl. Inf. Syst. **3**, 263–286 (2001). https://doi.org/10.1007/pl00011669
32. N. Marwan, How to avoid potential pitfalls in recurrence plot based data analysis. Int. J. Bifurc. Chaos **21**, 1003–1017 (2011). https://doi.org/10.1142/s0218127411029008
33. A. Zhang, Z. Lipton, M. Li, et al., Convolutional neural networks, in *Convolutional Neural Networks—Dive into Deep Learning 0.14.3 documentation* (n.d.). http://d2l.ai/chapter_convolutional-neural-networks/index.html. Accessed 31 Aug 2020
34. I. Goodfellow, Y. Bengio, A. Courville, Convolutional Networks, in *Deep Learning*. (The MIT Press, Cambridge, MA, 2017), pp. 330–371
35. S. Ioffe, C. Szegedy, Batch normalization: accelerating deep network training by reducing internal covariate shift **37**, 448–456 (2015)
36. N. Srivastava, G. Hinton, A. Krizhevsky, Dropout: a simple way to prevent neural networks from overfitting **15**(56), 1929–1958 (2014)
37. K. He, X. Zhang, S. Ren, J. Sun, Deep residual learning for image recognition, in *2016 IEEE Conference on Computer Vision and Pattern Recognition (CVPR)* (2016). https://doi.org/10.1109/cvpr.2016.90
38. L. Long, E. Shelhamer, T. Darrell, Fully convolutional networks for semantic segmentation, in *2015 IEEE Conference on Computer Vision and Pattern Recognition (CVPR)* (2015). https://doi.org/10.1109/cvpr.2015.7298965
39. I.F. Hassan, B. Lucas, G. Forestier, C. Pelletier, et al., InceptionTime: finding alexnet for time series classification (2019). arXiv:1909.04939
40. C.W. Tan, C. Bergmeir, F. Petitjean, G.I. Webb, Time series regression (2020). arXiv:2006.12672
41. J. Borkowski, Blocked designs (2014). http://www.math.montana.edu/jobo/thainp/ksampa.pdf. Accessed 31 Aug 2020
42. J. Demsar, Statistical comparisons of classifiers over multiple data sets. J. Mach. Learn. Res. **7**(Jan), 1–30 (2006)

Part V
Machine Learning in Video Coding and Information Extraction

Chapter 10
A Formal and Statistical AI Tool for Complex Human Activity Recognition

Anargyros Angeleas and Nikolaos Bourbakis

Abstract Recognition and understanding human activity is a challenging research topic for the last 30 years with significant advances, but with major open problems to be solved. Researchers have proposed a variety of advanced methodologies by approaching this interesting topic from a single direction. By saying a single direction we mean using either formal or statistical approaches. These approaches have their own limitations like the formal approaches cannot scale up beyond a certain level of complexity while the statistical ones don't work well in low complexities [1]. Thus, a possible solution is the use of a synergistic approach based on the efficient combination of formal and statistical methods. In this chapter, the authors offer a formal-statistical AI tool for the recognition of complex human activities. The formal model is associated with three context-free languages at body, motion, and activity levels. The statistical model is an artificial network that learns the activity of the human body.

10.1 Introduction

In the 1950's and 1960's, Alan Turing, Ross Quillian, Edward Feigenbaum, and Marvin Minsky were some of the very first notable pioneers who sought to replicate human-like intelligence for machines. As a direct result of their efforts, we are able to solve real-world problems today by experimenting with their approaches. Within this chapter, a novel end-to-end AI tool for complex human activity recognition and behavioral interpretation framework is presented, backed by formal and statistical information.

We start by offering a small introduction to the problem area as well as clearly defining the objectives of this tool.

Next, the formal and statistical approaches are presented along with the value of using both, how each one complements the other, and the work that is performed on

A. Angeleas · N. Bourbakis (✉)
CART Center, Wright State University, Dayton, OH, USA
e-mail: nikolaos.bourbakis@wright.edu

G. A. Tsihrintzis et al. (eds.), *Advances in Machine Learning/Deep Learning-based Technologies*, Learning and Analytics in Intelligent Systems 23,
https://doi.org/10.1007/978-3-030-76794-5_10

each step of the tool. Then, we present the architecture of the tool and implemented pipeline as well as graphical and statistical examples and information.

Finally, we wrap up the chapter via a small discussion about the learnings and other possible approaches that could tackle the identified issues.

Problem Area and Defining the Tool's Objectives

Human activity recognition is a well-studied area of research [2–6]. Despite recent advancements, monitoring systems still cannot function without human operators, which makes them inefficient. We know that the ability of humans to accurately identify behavior declines as the period of time completing the task increases, Kaber et al. [7].

It's obvious that in order to solve this problem, researchers must implement systems that not only "see and understand" the way a person does, but they must also be able to process raw streams of visual information and derive a result even more robustly than humans do.

Due to the wide variety of types of human activities, we first have to start from low-level actions and pose recognition. Then, we move toward systems that can use prior data and knowledge to more robustly recognize these actions and events [8–22].

Thus, the main part of this work is the classification of human activity for better recognition and understanding of human behavior, which is accomplished by efficient association of low-level and high-level reasoning. No matter how powerful and robust a system might be, activity recognition should be based on a set of rules that collaborate but do not limit one another [23].

In spite of the plethora of Single HAR approaches [24–27], we have not come across approaches that are able to provide spatiotemporal results and feedback for a sequence of complex activities. This part of the problem seems to be the least-researched area within the scientific community and has basically gone unexplored.

Thus, the main objective of this tool is to be able to identify complex human activities within controlled and uncontrolled environments and extract behavioral patterns derived from a rule-based system which is defined by domain experts. In order to do so, there are also important sub-objectives that need to be achieved.

To understand a simple activity, we need to implement the following actions: data collection, validation, manipulation, interpretation, and classification. Once we recognize a simple activity, we need to be able to understand whether the subject is still performing the same activity or if there is a change of activity.

To put a bit more structure to the aforementioned objective and sub-objectives, our approach to the activity recognition problem is based on the segmentation of the problem into three basic modalities:

1. Low-level: Pre-processing techniques that include but are not limited to video acquisition, background segmentation, and/or image denoising used to detect the points of interest.
2. Middle-level: Algorithms that track the subject within the sequence of video frames and extract features or contour.

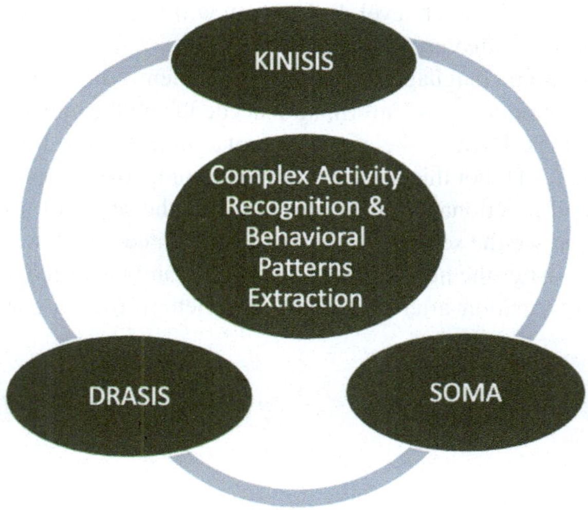

Fig. 10.1 The cooperative model for complex activity recognition

3. High-level: Algorithms that are used recognize *and* classify an activity. Some widely used techniques are: Hidden Markov Models, Hidden Random Conditional Fields, Artificial Neural Networks, and Deep Learning algorithms (Fig. 10.1).

10.2 The Hybrid Framework—Formal Languages

For the implementation of the Hybrid Framework, we utilize a combination of a Formal (Context-Free Formal Languages) and Statistical (Artificial Intelligence Algorithms) approaches [28–30].

Context-Free Formal Languages assert and communicate the approach as well as act as a validation mechanism of the data collection process along each step of the way. Utilizing Artificial Intelligence Algorithms allows for scalability, robustness, ease of implementation, and agnostic nature when it comes to solving a problem.

Coupling these modalities provides the tool with all needed contextual information within each frame. It gives us the ability to symbolically represent the activities that stem from the recognition engine while enhancing our tool to extract grammatical, visual, and statistical results.

As it has been previously explained, we define and divide Human Activity Recognition into three modalities: Low-, Middle-, and High-level. Thus, we use a collaborative three formal-languages model. Each modality can communicate with one another, but they can also work independently and, if manipulated, uniquely approach the HAR problem.

We named the Framework SDK (*SOMA-KINISIS-DRASIS*). Soma is a Greek word that means body; *KINISIS* is a Greek word that means movement; and *DRASIS* is another Greek word that means action. Each name succinctly describes what happens at each level. Below, you will find a bottom-up visual representation of the tool architecture (Fig. 10.2).

Soma language is responsible for identifying, tracking, extracting, validating and manipulating the information needed from the human subject. Identifying, tracking, and performing skeletonization of a subject is a well-versed and well-researched area [27, 31]. For this reason, we are not going to dive deep into presenting the algorithms and functionality, as they are outside the scope of this work. Within Fig. 10.3 you can see the selected body parts highlighted—yellow.

After the human subject is tracked and identified, our tool extracts the necessary information after applying the skeletonization algorithm to the subject. The tool

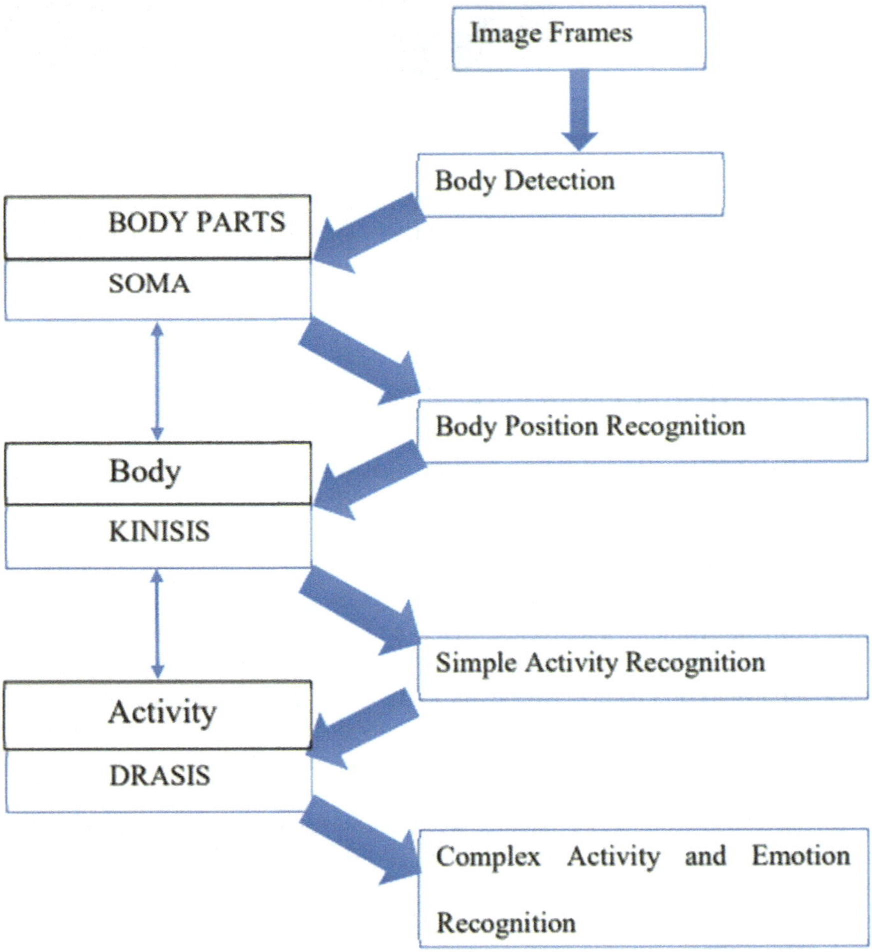

Fig. 10.2 The three-layer formal model for complex activity recognition

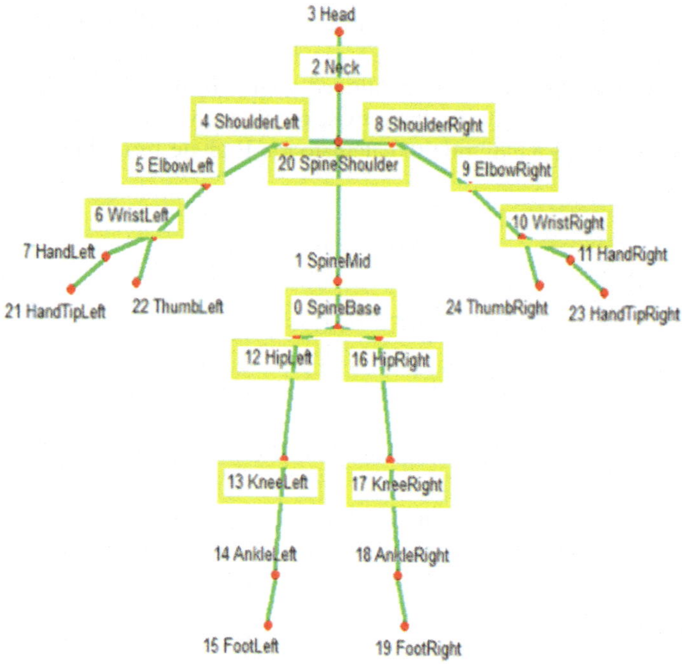

Fig. 10.3 The selected human body parts

extracts the x, y, z axis information for the following body parts while also calculating the acceleration of the torso of the human subject—a(t):

- Neck
- Spine shoulder
- Spine base
- Left shoulder
- Left elbow
- Left wrist
- Left hip
- Left knee
- Right shoulder
- Right elbow
- Right wrist
- Right hip
- Right knee.

After running various experiments, we decided to utilize the minimum amount of modalities that output acceptable recognition outcomes, given that one of our goals is to create a robust tool while maintaining a low computational cost.

The next step is an important contribution of this tool. After we are aware of the position in the spectrum of the body parts, we calculate the angle created in between the adjacent body parts mentioned in the list above, the inclination angle with the ground of the torso, upper and lower leg, as well as the acceleration of the torso.

In order to calculate the aforementioned angles in the 3D space, we use the following equations. (Of course, there exist numerous ways to do this, but we use vectors and "the right-hand" rule to calculate the B angle from three points: A, B, and C.)

$$AB^{\rightarrow} = B - A \tag{10.1}$$

$$BC^{\rightarrow} = C - B \tag{10.2}$$

The scalar product or dot product has the following property.

$$AB^{\rightarrow} * BC^{\rightarrow} = ||AB^{\rightarrow}|| * ||BC^{\rightarrow}|| * \cos\cos\theta \tag{10.3}$$

With ||*||, we measure the length, and θ is the angle formed among the two vectors. Thus, all the information we need are the (x, y, z) coordinates of the three points, and then we can rearrange Eq. 10.3 and solve it for θ.

$$\theta = ar\cos\left(\frac{AB^{\rightarrow} * BC^{\rightarrow}}{||AB^{\rightarrow}|| * ||BC^{\rightarrow}||}\right) \tag{10.4}$$

In order to calculate the acceleration of the torso, and given that we know its position in the 3D space, we need to do the following:

We know that given the position function of an object, the velocity of the object is the first derivative of the position function, while the second derivative of the position function, or the first derivative of the velocity function, produce the acceleration of the object.

So, we can calculate the acceleration of the object using the following formula:

$$a(t) = \frac{v(t + \Delta t) - v(t)}{\Delta t} = \frac{dv}{dt^2} = \frac{dv_x}{dt} \tag{10.5}$$

After the aforementioned data points are compiled, the data points are passed to KINISIS language. Here, we should mention that we utilize a continuous data stream. The main benefit of this is data streams can be processed, stored, analyzed, and acted-upon as they are generated in real-time. Also, the inclination angle in-between the torso, knee, and the ground are parsed by DRASIS in order to calculate the change of activity.

The KINISIS language objective is to classify action data into simple activities. For example, walking is a simple activity that consists of the actions of moving one's legs and hands. SOMA language continually parses and streams the selected adjacent body part (actions) data, while KINISIS constitutes a sequence of time series information with the goal of classifying these activities.

A preview of KINISIS Language output for a recognize activity

Bst# → BaHRU # BaHLU # BaFRU # BaFLU # H(RU)aH(RL) # H(LU)aH(LL) # F(RU)aF(RL) # F(RU)aF(LL)

Σ Bst -> {Bst0, Bst1, Bst2, Bst3,....., Bstn, Bstn+1} -> Walking

Fig. 10.4 An example of frame data and a sequence of frame data

The first line on the Fig. 10.4 showcases the angles generated among adjacent body parts for a selected frame. The second line presents a sequence of information extracted from frames that were classified as walking.

An important contribution of this approach is the ability to reduce the frame which needs to be parsed. According to the benchmarks run with our machine learning model, we identified there is not a significant increase when all the output frames are utilized per second.

Our hypothesis is based on heuristics and according to IBM's multimedia parsing study [32]. To further validate our approach, we ran benchmarks for 30 frames per second (f/s), 10 f/s, 5 f/s, and 3 f/s, and the activity recognition rate changes were miniscule—between 30 f/s, 10 f/s, 5 f/s. Because of this, and given our goal of developing a robust end-to-end tool, we decided to utilize the 5 frames per second approach.

Up until now, we have defined the way we recognize actions (SOMA) and classify activity (KINISIS), but the end goal of this tool is to be able to recognize sequences of complex activities. Therefore, our next step is to present DRASIS formal language.

The role of DRASIS is to recognize whether or not a change of activity exists and to extract behavioral assumptions derived from a rule-based system. For the purpose of this study, change of activity is defined as the fuzzy moment where a person transitions from one activity to another.

According to literature, the vast majority of human activity recognition systems recognize and classify change of activity at one single point [33–36]. The limitation of these studies is that they cannot recognize the temporal description of each approach. The reason we utilize a fuzzy approach is because every single person transitions from one activity to another differently. Additionally, the temporal information is very important to be able to extract behavioral patterns.

For the action segmentation section, we will use three (3) angles that are generated from the skeletonization of the following three components: body, upper legs and lower legs, and acceleration of the torso. We should note that we will not use the angles that are generated from the relation among them, but the inclination angle of these three body parts in relation to the ground. This information is going to be parsed directly from the SOMA language.

In conclusion, our recognition tool works as follows: SOMA language is a precursor of KINISIS language, although one is not able to produce without the contribution of the other. SOMA language is responsible for extracting and identifying unique characteristics of a human body in a specific video frame. KINISIS language receives this information and unifies it in pursuance of simple activity. And finally, DRASIS language receives simple activities, determines the turning points, and then recognizes the complex activity.

The reasons for incorporating a statistical approach for the recognition aspect of KINISIS and DRASIS languages:

1. Grammars expect high accuracy in the lower levels, hence they are not suited to deal with errors in low-level tasks such as tracking errors and missing observations. (Incorporation of ANNs for Activity, and Change of Activity Recognition)
2. Formulating the rules for scenarios far more complex than a sequence of actions increases, like in overlap and parallelism of actions. (Coupling ANNs, Fuzzy Cognitive Maps, and Rule-Based Colored Petri Nets)
3. Using grammars for training and learning is an intelligent alternative but very computationally intense and not at all robust. (Learning is performed with ANNs)
4. Formal Languages are not adequate enough for expressing statistical results and scaling. (Statistical are given with the incorporation of ANNs and scaling results represented with Rule Based Colored PNs)

10.3 Formal Tool and Statistical Pipeline Architecture

The high-level overview of the system's functionality is described within the following 11 steps and depicted in the image afterward:

1. KINECT camera starts recording.
2. SOMA begins.
3. Pre-processing steps include but are not limited to tracking, motion detection, background subtraction, silhouette extraction.
4. Body skeletonization with the use of KINECT library.
5. Parse and store body joints positions in time (4D space).
6. KINISIS begins.
7. Calculate eight (8) angles among adjacent body parts and acceleration of the torso in the x-axis.
8. Manipulate data and create nine (9) feature vectors.
9. With the use of Artificial Neural Networks, the activity is recognized within the first minute of the video.
10. DRASIS begins.
11. Dual role:

    a. Recognize change of activity or absence of change of activity.
    b. Classify a sequence of activities as normal or abnormal.

Our system's function is depicted in the Fig. 10.5. Blue modalities describe SOMA language, orange modalities describe KINISIS language, and yellow modalities depict DRASIS language.

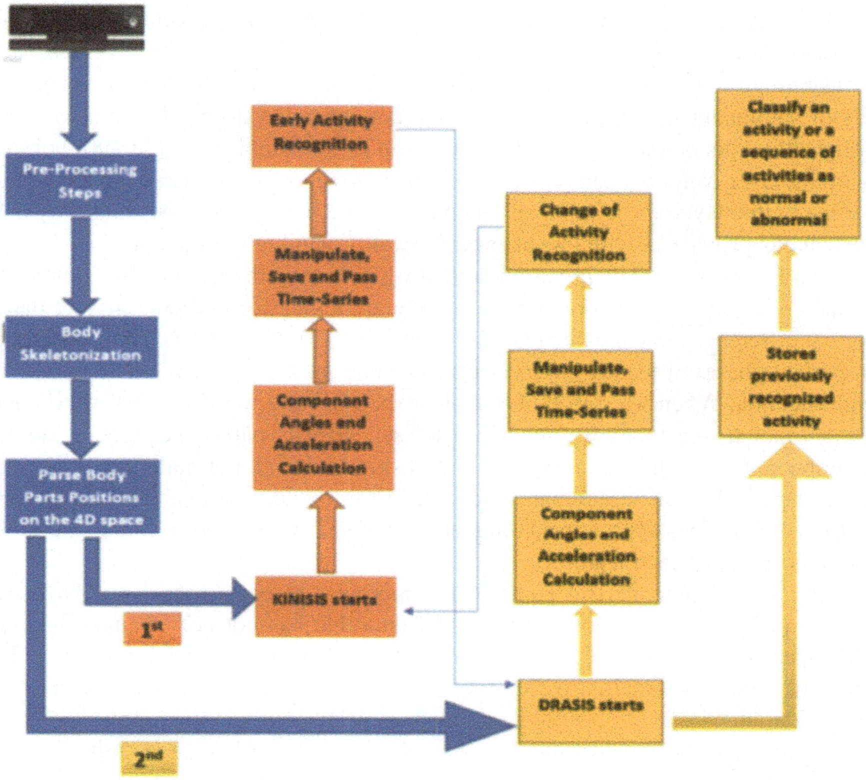

Fig. 10.5 The view of the tool's functions

10.4 DATA Pipeline

The main objectives when it comes to the creation of an end-to-end AI pipeline are to: reduce latency, create loosely coupled modalities, ensure the architecture is horizontally and vertically scalable, generate efficiency relative to workload management, and be fault-tolerant.

Our data pipeline is comprised of two well-defined components: online model analytics, where the model is applied for new data decision-making, and offline data discovery, where the system learns based on historical data to create the AI model to be used in batch-processing mode.

As discussed prior, one of the most vital aspects of an AI tool is the data pipeline, and this anecdotal quote helps us understand why: “Data scientists spend 80% of their time cleaning data rather than creating insights. Data scientists only spend 20% of their time creating insights; the rest wrangling data.” It’s frequently used to highlight the need to address a number of issues around data quality, standards, access.

Within the next few paragraphs, we will describe eight (8) important elements of an AI pipeline.

Data Ingestion: Acquiring and storing data is the first step of any AI workflow. It's of significant importance to ensure the data acquired do not get transformed, which allows us to have the collected dataset in an immutable state [37, 38]. For the learning layer (offline), you can apply numerous techniques to parse the data, but, for the online layer, a streaming architecture should be implemented. This way our system can provide us with low latency, reliability, and high throughput.

Data Preparation: Data preparation involves assessing the validity and condition of the collected data. Important actions include looking for outliers, missing data, format difference, and any other type of anomalies [39, 40]. Here is where the three important phases of feature engineering come into play: extraction, transformation, and selection. When we start building and experimenting with Datasets, we do not know which features we are going to use beforehand. This part of the pipeline should be implemented with re-usability and composition in mind. For both the offline and online layers, this can be automated if domain expertise does not exist. We can always utilize statistical tests, principal component analysis, or any other technique to rank the importance of each feature.

Data Segregation: It is of paramount importance to use a model that is based on the quality of its pattern prediction for data that has not been trained on. For this given problem, we use the standard protocol defined for neural networks. We randomly slice our data to use 70% for training, 10% for testing, and the other 20% for validation cases.

Candidate Model Evaluation: This step occurs on the offline layer, and it can be scheduled based on how critical the application is and how often it needs to be used. Here, we should always remember to implement error tolerance, data checkpoints, and failover enablement for data partitions. We will go into more detail regarding the model training service within the next section, where we present the best model for the HAR and change of activity detection. We should always keep in mind that the more data we get, the higher the probability our model will change. Therefore, this service needs to be able to automate this process, run various experiments, and call different AI algorithms.

Model Deployment: For real-world applications, once there is a winner, model-wise, it is typical for the model to be deployed and applied to the decision-making framework.

Model Scoring: This is the step where we identify whether the selected model performs well or not. For this given problem, we utilize the absolute recognition rate on our validation dataset, and, for the offline experimentation, we utilize the recognition rate and the confusion matrix to understand and improve the feature selection, model type, and hyper-parameters.

To recap, always start with identifying the problem and making assumptions regarding which types of features should be tested to validate or invalidate a hypothesis (data engineering). Then, ideate and specify the features to be used as well as the way the data should be collected. Finally, create the data pipeline that correlates to the above steps—one that is specific, but broad enough to allow experimentation with different AI algorithms.

10.5 Tools for Implementation

Within this section, we are going to present steps taken to implement the HAR tool, the hardware, and software technologies used to implement the HAR tool. The programming languages used are:

- C++ for the configuration and manipulation of the KINECT Camera
- C++ for the data preparation and validation parsed from the live feed
- Matlab for rapid experimentation and prototyping for the AI model
- Python for the AI Pipeline.

For the live data collection part of the tool, i.e. recording human subjects to perform activities, we utilize Microsoft KINECT V2. We do so because KINECT is widely used in the field of computer vision for human activity recognition, gesture recognition, activity analysis, object recognition and tracking, etc. When we compare KINECT with conventional 3D cameras, like stereo cameras and range cameras, there are obvious advancements and enhancements it offers. A thorough description of these is provided by Gokturk et al. [41] (Fig. 10.6).

Graphical User Interface for Real-Life Data Collection
The Graphical User Interface consists of three modules. The Camera module is used to take snapshots of users via the RGB camera, the infrared sensor, and the depth sensor. Skeletonization is applied to the pictures as well. The Recording module creates a video of everything that can be seen by the KINECT device. Finally, the

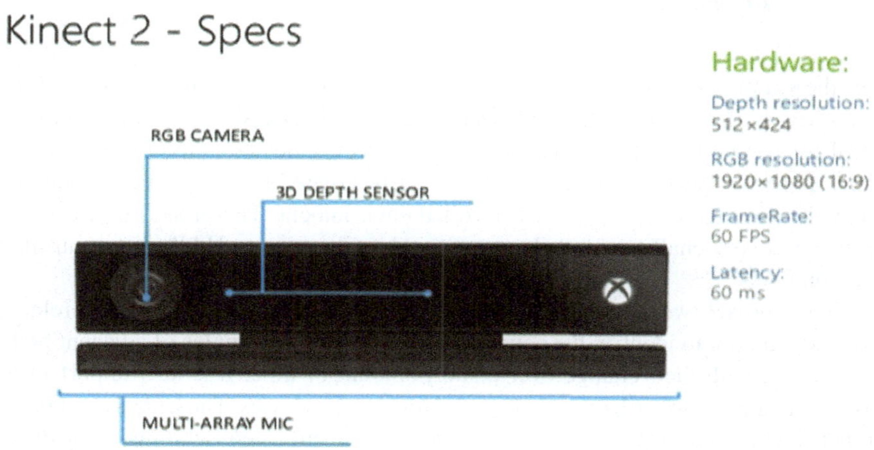

Fig. 10.6 The Kinect V2

Angles module depicts the recordings of the RGB camera and shows information for the angles generated among the adjacent body parts.

When we are in this module and choose to record, the video stream is saved directly to the pictures folder, and a text file with all the angles recorded in each frame is generated and saved in the Documents folder. The GUI is implemented in C++ utilizing the KINECT and OpenCV libraries.

The process of body identification within a sequence of frames is the following:

- Determine the reference point of the body frame.
- Check if the body frame is null.
- Initialize a body list to store the joints that the sensor detects.
- Obtain and refresh the body data for every new frame.
- Loop through the list of bodies if there are more than one.

10.6 Experimentation with Datasets to Identify the Ideal Model

We will start with the way we approach the modeling of single human activity recognition, then move toward the way we approach change of activity recognition. Next, we will present the implemented GUI responsible for data collection when it comes to real-life human activity recognition. Finally, we will present a rule-based model that extends the functionality of our tool to become a domain-specific application. For example, a geriatric doctor can utilize the tool to track patient-specific data and extract behavioral conclusions.

10.6.1 KINISIS—Single Human Activity Recognition Modeling

For the simple human activity recognition portion of the tool (for which KINISIS is responsible), we utilize the MSR Daily Activity 3D Dataset [42]. This Dataset is ideal for our tool for three reasons. First, a KINECT device is used to record the daily activities. Second, the length of the videos created is very short, and, finally, the videos are recorded in a non-controlled environment. Therefore, the Dataset is considered very challenging and a great fit when it comes to HAR in uncontrolled environments.

The approach we take for recognizing single activities is a time-series-related one. We attempt to identify the activity based on a batch of angles of adjacent body parts along with their change over a short amount of time. It is also important to communicate that we utilize multi-featured time-series entries. Thus, ideal candidates to benchmark and identify the best model to utilize are Deep Learning algorithms. And, more specifically, Convolutional Neural Network (CNN), Long Short-Term

Neural Network (LSTM), and a hybrid model of these two are used to classify the activity. The reason we utilize these three models for our experiments is partially because of theoretical defined benefits of those networks as well as measurably quantifiable results from other studies, including but not limited to:

- Panzner et al. [43] conducted research regarding the applicability of LSTM and HMM models in activity recognition.
- While Fazle et al. [44] performed a study on Multivariate LSTM and FCN for Time Series Classification.

To fine-tune the models and obtain the best possible results, we implement Exhaustive Grid Search [45]. Furthermore, when identifying the parameters of a prediction model, it is tremendously incorrect to use the same labels and samples, as this will cause the model to be unable to predict unseen data with high accuracy. This phenomenon is called overfitting [46]. Best-practice to tackle the overfitting problem when it comes to supervised learning is to always utilize a random part of the available data. Below is a flowchart that showcases how to implement cross-validation to avoid the overfitting problem as well as exhaustive grid search to finely tune the model's parameters, Fig. 10.7.

To review, the process of identifying and finely tuning the model is as follows:

- Identify the type of data that is going to be used.
- Research and try a number of models that can tackle the problem.
- Further understand the models by investigating their pros and cons, and select the most appropriate for the given situation.
- Define the model validation approach (e.g. we selected segmenting our data into 70% training data, 10% test data, 20% validation data).

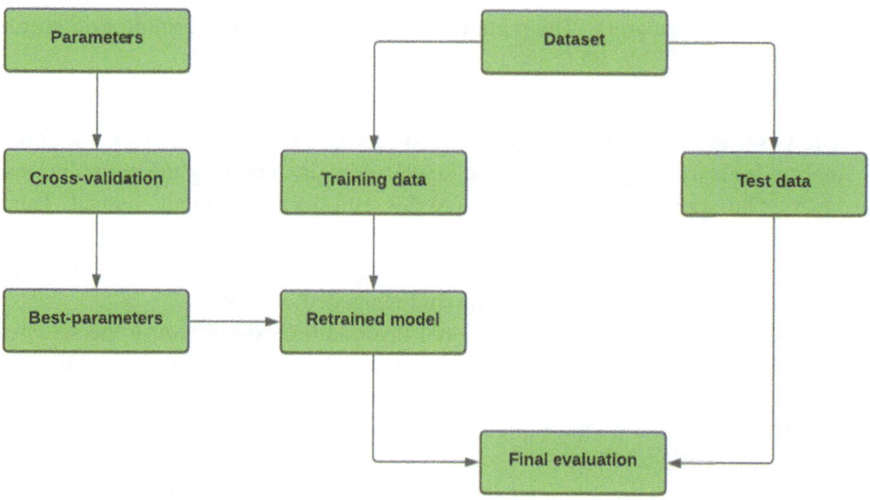

Fig. 10.7 Data pipeline flowchart

Table 10.1 CNN optimal architecture

Convolutional neural network	
Number of convolutional layers	4
Second, third, and fourth convolutional layer (activation function/pooling layer function)	Hyperbolic Tangent function (tanh)/max pooling
First, and fifth convolutional layer (activation function/pooling layer function)	Softplus function (relu)/max pooling
Learning rate	0.001 (default)
Batch size	300
Sequence length	128
Epochs	200
Cost function	Softmax Cross Entropy with Logits v2
Cost optimizer	Adam Optimizer
Accuracy	Reduce mean (keeps a running average of the results of computation from a batch of inputs)

– Utilize grid search or a respective methodology to experiment with your dataset.

Moving forward, we will present the models and the recognition accuracy results and compare them with other methodologies/approaches.

The following resources are very valuable for learning more about CNN, LSTM, and CNN-LSTM models [47–54].

Convolutional Neural Networks

First, we will provide the information regarding the architecture of the network that delivers the best results. Then, we will present the evaluation results on the set as well as the overall test accuracy. The parameters identified from performing grid search are depicted in the Table 10.1.

Some interesting findings include:

– The CNN model outperforms the LSTM and CNN-LSTM. CNNs parse information both from the temporal and spatial domains utilizing the convolutional layers. Our approach parses the time-series data from each input channel. Then, convolution and subsampling are performed, and, as a result, numerous channels of information are generated from neighboring input time-series data. Finally, all data information is combined to present the needed outcome.
– We identified that six (6) convolutional layers are needed because the selected depth is adequate enough to produce meaningful and accurate results. We could add more layers, but this won't enhance the activity recognition accuracy by more than 0.832%. When we have three or four convolutional layers, the accuracy rate drops from 2.192 to 4.73%.
– A batch size much greater or smaller than six hundred (600) hinders accuracy. To be precise, a batch size smaller than four hundred and fifty (450) or greater than six hundred and fifty (650) lowers the accuracy from 0.342 to 3.927%.

- The activation functions for the convolutional layers also play a big part. We found that no matter the network's architecture, we always manage to achieve the best performance when the first and last convolutional layers use the softmax function. For hidden layers, the hyperbolic tangent function was identified as the best performing one. That's another heuristic denouement after performing various benchmarks using different combinations of functions.
- The second version of the softmax cross entropy with logits is used for the cost function, and the Adam Optimizer is used for the cost optimization function since its goal is to minimize the cost of learning.

The overall Test Accuracy is 95.307%, coming first within the three networks.

Long-Short Memory Neural Network Architecture

We will follow the same architecture for communication findings for the LSTM network as we did for the Convolutional Neural Network. First, we will provide the architecture information of the network that delivers the best results. Then, we will present the evaluation results of the set as well as the overall test accuracy. The parameters identified from performing grid search are depicted in the table below, Table 10.2.

Some interesting findings include:

- We expected that more hidden layers would enhance the activity recognition, but the accuracy rate actually dropped. For example, when we added a third layer, the accuracy dropped by 2.083% when all other parameters were unmodified.
- As mentioned earlier, LSTM networks require relatively small batch sizes in comparison to other neural networks, like CNNs. Our network produced almost identical results with batch sizes of one (100), two (200), or three hundred (300). We chose to use the latter because the network slows with smaller batch size.

Table 10.2 LSTM optimal architecture

Long short-term memory neural network	
LSTM size	27 (three times the input channels, default approach)
LSTM layers	2
Batch size	300
Sequence length	128
Epochs	200
Activation function	Softplus function (relu)
Cost function	Softmax cross entropy with Logits v2
Cost optimizer	Adam optimizer
Gradient clipping	Adam optimizer
Dropout parameters	Training: 0.25 Test: 1.00
Accuracy	Reduce mean (keeps a running average of the results of computation from a batch of inputs)

- The activation function plays an equally important role. We found that regardless of the network's architecture, we managed to achieve the best performance when Softplus (ReLu) was the activation function. This is another heuristic denouement we identified after performing grid search.
- After carefully reviewing the literature and research studies [55–58], we realized we needed to select the second version of the softmax cross entropy with logits as our cost function and use the Adam Optimizer for the cost optimization function.
- Gradient clipping technique impeded the problem of exploding gradients which occur in deep neural networks. This problem is more prevalent for Recurrent Neural Networks, like the LSMT. There exist numerous gradient clipping techniques, but we selected the most widely used function, Adam Optimizer.

Finally, the overall Test Accuracy is 92.959%, ranking second among the three networks.

Convolutional Long Short-Term Memory Neural Network

Before we performed benchmarks, we expected that the CNN-LSTM model would yield the best results because it encompasses the best attributes of both worlds. This type of network naturally incorporates the time-series vectors, which allows the network to quickly recognize and efficiently interpret ambiguity. Within our experiments, we provide each activity's time-series as a single vector with fixed length, Table 10.3.

Here are some interesting findings:

- When we add more convolutional layers, or even a pooling layer, the performance of the networks drops drastically.

Table 10.3 C-LSTM neural network optimal architecture

Convolutional—long short-term memory	
Neural network number of convolutional layers	1
Convolutional layer (activation function/pooling layer function)	Softplus function (relu)/max pooling
LSTM size	27 (default: three times the amount of channels)
LSTM layers	2
Learning rate	0.001 (default)
Batch size	300
Sequence length	128
Epochs	200
Cost function	Softmax cross entropy with logits
Cost optimizer	Adam optimizer
Accuracy	Reduce mean (keeps a running average of the results of computation from a batch of inputs)

Fig. 10.8 Real-time data collection and interpretation

- The above hyperparameters were identified as the best performing after conducting various experiments with grid search.

The overall Test Accuracy is 88.803%, ranking second among the three networks.

Below is an illustrative example of a subject walking within a controlled environment recorded using our tool, Fig. 10.8.

SOMA collected the information and interpreted the data.

Bst1 → {BaHRU(121) # BaHLU(128) # BaFRU(71) # BaFLU(84) # H(RU)aH(RL)(200) # H(LU)aH(LL)(186) # F(RU)aF(RL)(198) # F(RU)aF(LL)(151)}

Bst2 → {BaHRU(140) # BaHLU(177) # BaFRU(138) # BaFLU(67) # H(RU)aH(RL)(190) # H(LU)aH(LL)(178) # F(RU)aF(RL)(193) # F(RU)aF(LL)(151)}

Bst3 → {BaHRU(146) # BaHLU(166) # BaFRU(166) # BaFLU(106) # H(RU)aH(RL)(189) # H(LU)aH(LL)(177) # F(RU)aF(RL)(204) # F(RU)aF(LL)(153)}

Bst3 → {BaHRU(121) # BaHLU(128) # BaFRU(71) # BaFLU(84) # H(RU)aH(RL)(200) # H(LU)aH(LL)(186) # F(RU)aF(RL)(198) # F(RU)aF(LL)(151)}

Bst4 → {BaHRU(140) # BaHLU(177) # BaFRU(138) # BaFLU(67) # H(RU)aH(RL)(190) # H(LU)aH(LL)(178) # F(RU)aF(RL)(193) # F(RU)aF(LL)(151)}

Table 10.4 Comparing against state-of-the-art approaches

Proposed approach	Accuracy %
Celiktutan et al. (2015)	72.90
Azary and Savakis (2013)	78.50
Enea Cippitelli et al. (2016)	81.20
Azary and Savakis (2012)	83.90
Our approach (CNN-LSTM)	88.03
Chaaraoui et al. (2013)	90.60
Zhu et al. (2013)	90.90
Zanfir et al. (2014)	91.70
Vemulapalli et al. (2014)	92.46
Our approach (LSTM)	92.95
Chaaraoui et al. (2014)	93.50
Li et al. (2010)	94.20
Our approach (CNN)	94.47
Yong Du et al. (2015)	94.49
Xia et al. (2012)	97.15

Bst5 → {BaHRU(146) # BaHLU(166) # BaFRU(166) # BaFLU(106) # H(RU)aH(RL)(189) # H(LU)aH(LL)(177) # F(RU)aF(RL)(204) # F(RU)aF(LL)(153)}.

This is a sample of the movement of the subject. The data parsed by KINISIS, and the activity was classified as walking.

Within KINISIS implementation section we tried and achieved with a good accuracy to classify ongoing videos for human activity recognition. LSTM, CNN, and CNN-LSTM networks are widely used for activity classification. We ran benchmarks using these three types of networks, we tuned the parameters, and we changed the shape of the network, in pursuance of a high recognition rate for activity classification.

Finally, and before we move on to the next part of this tool, that is the temporal understanding of change of activity, we are going to compare our approach with state-of-the art results. Reference, Table 10.4.

10.6.2 DRASIS—Change of Human Activity Recognition Modeling

DRASIS is the high-level knowledge extraction modality of this tool which parses body state information directly from SOMA. As mentioned earlier, there is not abundant research for the change of activity recognition problem, and the approaches taken do not identify the temporal element of the change of activity, which is very important to our tool [59–62]. In order to experiment and determine a way to approach

Table 10.5 DRASIS—LSTM optimal architecture

Long short-term memory neural network	
ISTM size	15 (three times the input channels, default approach)
ISTM layers	1
Batch size	23
Learning rate	0.00001
Sequence length	10
Epochs	50
Activation function	Rectified linear unit (relu)
Cost function	Softmax cross entropy with logits v2
Cost optimizer	Adam optimizer
Gradient clipping	Adam optimizer
Dropout parameters	Training: 0.25 Test: 1.00
Accuracy	Reduce mean (keeps a running average of the results of computation from a batch of inputs)

the problem, while being able to recognize spatiotemporal data, we utilize the UTK KINECT—Action 3D Dataset [63].

We approach the change of activity problem as binary multivariate time-series. Because the data we have are not enough to experiment with CNN, which yields the best results for single activity recognition, we utilize only the LSTM, which learns/converges faster, Table 10.5.

We finely tuned the LSTM model hyper-parameters using grid-search and ran the experiments using 80% of the data for training and 20% for testing. We changed our strategy here because the dataset is not comprehensive enough to allow us to go with a 70–10–20% approach.

Finally, the overall Test Accuracy is 81.037%. We firmly believe that we would be able to achieve a higher recognition rate with the use of a more comprehensive Dataset.

Now, utilizing the domain experts' knowledge, we create a rule-based system to identify behavioral patterns using extracted spatial and temporal data for a sequence of activities.

In order to extract behavioral patterns, we need to be able to create rules, apply rules, and create a scalable tool. We can graphically represent the knowledge of the domain expert utilizing Dynamic Fuzzy Cognitive Maps (DFCM). Michael Glykas's book [64] perfectly depicts the usefulness and applicability of FCM in modern systems. Kasper Kok's et al. [65] paper reviews the growing studies of FCMs and DFCMs. Essentially, the only difference between FCMs and DFCMs is that the latter may evolve based upon new observations and create new relations among nodes.

From now on and whenever we use FCM, we define C as the fuzzy set, where $ci \in C$ and W measures the strength of the relationships between the elements of C. We use the FCM because we want to optimize the learning process, for the associations among activities. During the construction of the FCM the characteristics of the activities are taken under consideration, as well as the temporal and local characteristics of them.

After identifying the concepts, and the relationships among them, the weighted connections are determined. We characterize every activity in terms of the direction of the relationship, either direct, or inverse, and a qualitative assessment of the magnitude, as well as the bi-directional or non-existent relationship. Another learning modality that can be extracted via an FCM is the degree of affections among modalities. Meaning that if a concept changes how much is going to affect the interconnected concepts, or how and if will it make new relationships with concepts that are not connected by then.

All cohort groups are modeled independently and that allows us to rehash causal factors, omitting factors that we consider as not irrelevant for each group and adjust the relationships between factors to portray the analysis group by group. In all our examples weights are equally distributed but in a real-life system, these weights are going to be adjusted according to prior observations, and the probabilistic element of which activity might show up after the previous one will be further enhanced to the system.

That is going also to help us with regard to the classification of the next activity. Hence, we are going to be able to eliminate action classes that cannot be performed after the last activity detected.

Within the next few pages, we are going to present an example of a behavior extraction system that detects normal and abnormal behavioral sequences. Below, we can view the possible activities that relate to one another for normal behavioral sequencing from the UTK Kinect—3D Action Dataset, Table 10.6.

And these can be translated to a DFCM graph like the one below, Fig. 10.9.

Finally, to interpret the sequence of activities as normal or abnormal, we utilized Rule-Based Colored Petri Nets. Petri Nets were used because they visually depict the information; their tokens give us the advantage to replicate the concurrent and dynamic nature of activities; and they offer timing capabilities for the execution of an event, top-down and bottom-up hierarchical modeling, abstracted processes, and high-level reasoning.

Depicted below is a Rule-Based Colored Petri Net, Fig. 10.10.

Table 10.6 Fuzzy set modalities and accepted activities per modality

Fuzzy set modalities	Activities relations
C1: sit–down	{C3, C5}
C2: run	{C1, C2, C4}
C3: lay	{C3, C5}
C4: walk	{C1, C2, C4, C6}
C5: stand–up	{C1, C2, C4, C6}
C6: grasp and carry	{C1, C2, C4, C6}

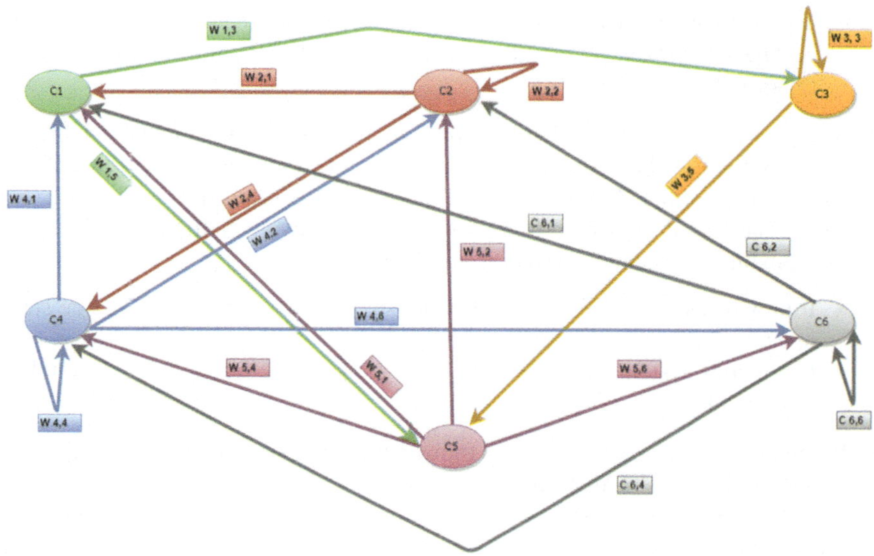

Fig. 10.9 The DFCM diagram

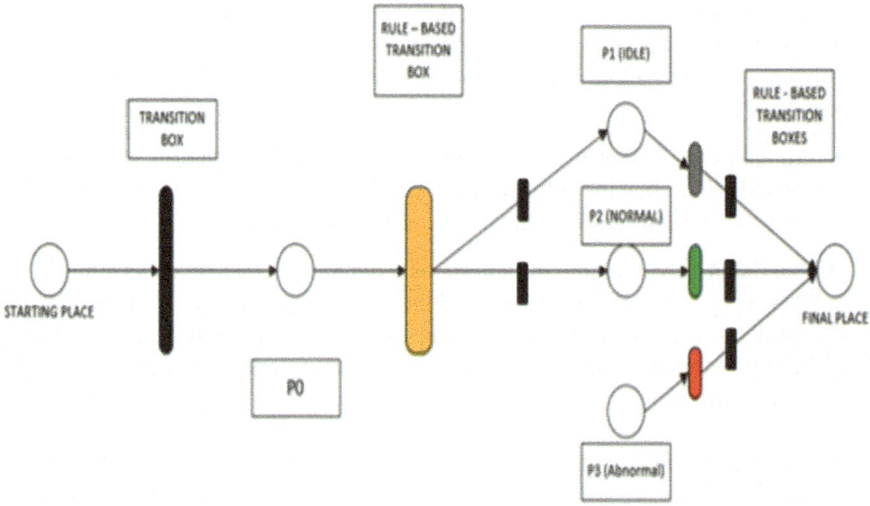

Fig. 10.10 A rule-based colored petri-net

- The first black transition box is an immediately activated transition box when a token is generated.
- Then, the token stays in place (PO), and a yellow Rule-Based transition box is implemented.

  - This box categorizes the token as idle or normal. When the system can't detect a subject, the token goes to idle. When an activity is recognized, it always goes to normal.

- The token placed in P1 will always be colored grey. The grey token depicts the idle condition of the system. In order for this token to move to the Final Place, the condition of the Grey Rule-Based Transition Box needs to be satisfied.
- Grey Rule-Based Transition Box will always have two conditions, regardless of the domain for which the system is going to be used.
  - If previous system state == "idle" and subject ! = "recognized", token stays in P1
  - If previous system state == "idle" and subject == "recognized", the token moves to Final Place
- When an activity is recognized, it follows the same steps as the grey token but instead will always be in place P2.
- P2 is when our system is in normal condition, i.e. when the detected activity or sequence of activities are normal. We should note here that there is a chance that a new activity might make our system move to an abnormal state if a rule is fired.
- P3 represents the system's abnormal place. In order for the system to be in an abnormal place, the Green Rule-Based Transition box should fire, and instead of moving the tokens to the final place, they are moved to place P3.
- The Red Rule-Based Transition box moves tokens to the final place. When this happens, an alert will be generated because something abnormal happened to the subject.
  - Another case is that the subject indeed performed an abnormal sequence of activities, but then the subject transitioned to a normal activity. So, we change the system's state in Normal, and move tokens back to P2.

Within the first three graphs system sends a signal that it is active, but still there is no subject recognized, after doing so, and the token stays at place P1 of the RBCPN graph, until a subject is recognized. The idle token appears fist at the "starting place", passes the first transition box, and stays momentarily in place P1, then passes the Rule-Based Transition box, and according to the rules that it encompasses is going to be categorized, as idle system, or normal activity. Obviously when the system informs us that it is idle, then the token is going to be sent to the idle place P1. Within the next example we are going to describe how a detected activity moves within the RBCPN graph, Figs. 10.11 and 10.12.

During timestamp t1 our system informed us that it is in an idle state, then during timestamp t2 KINISIS classified an activity as walking, and DRASIS. Then the RBCPN modality of our system takes charge. So, after passing from the starting place, moving forward to place P1 and getting through the first transition box, walking token is in place P2, obviously we regard walking activity as a normal one. Before we move forward we should note here that if there are no previous observations, all first recognized activities are going to be categorized as normal. Actually, a recognized

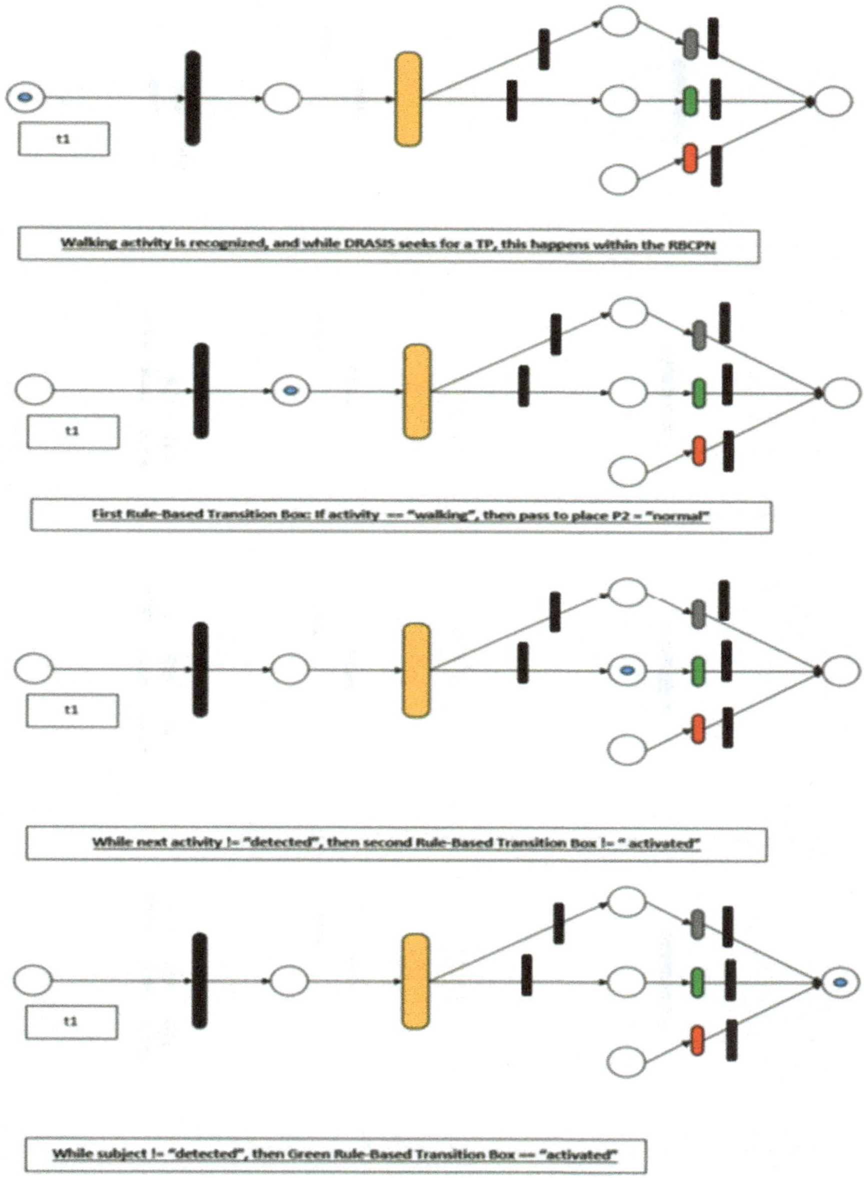

Fig. 10.11 A rule-based colored petri-net depicting different activation states (**a**)

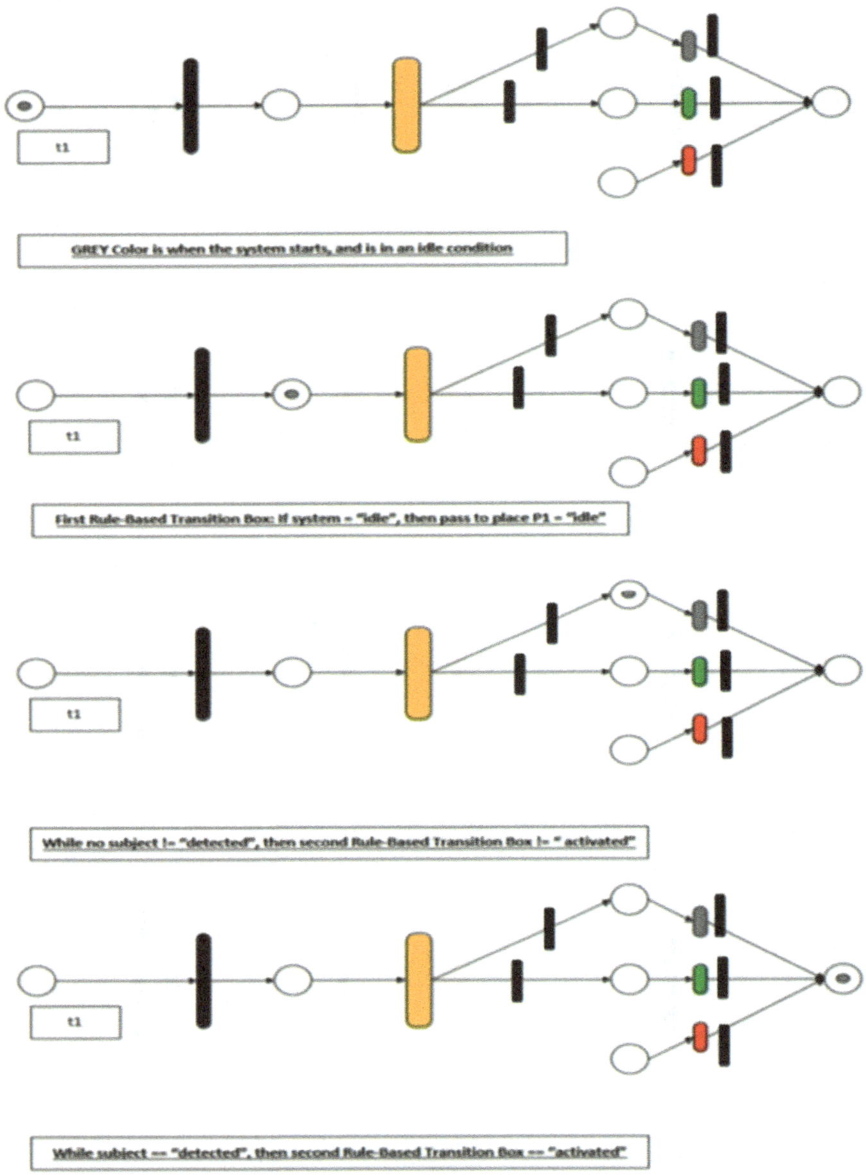

Fig. 10.12 A rule-based colored petri-net depicting different activation states (**b**)

activity is always going to be categorized as normal. Then based on our rules if the system won't recognize new activities, and based on the activity that is performed that state of the system might change. The Rule-Based transition boxes, when an activity passes this special case transition boxes, the system checks the rules that we set within each one of them. We parse the information regarding activity connectivity from the FCM, as well as from other rules than we add. But the rules should not only include acceptable transitions, but also what we consider not-acceptable transitions, as well as normal/abnormal transitions. Based on this functionality, and according to the area that the system is going to be used what is normal/abnormal, acceptable/not-acceptable might change.

Utilizing the above described Rule-Based Petri-Net and the reasonably sound rules along with the activity recognition part of the system, we extract normal and abnormal behavioral patterns.

For example:

- Sit-Down & Lay → If two time-stamps passes, and activity = "lay", call emergency
- Walk & Run & Walk & Run & Walk & Sit-down → If twenty time-stamps passes, and activity = "Sit-down", call emergency.
- Walk & Run & Walk & Run → working out Walk & Grasp-Carry & Leave on Desk & Walk & Grasp-Carry & Through → Performing house work
- Walk & Sit-down & Stand-Up & Walk & Sit-down & Stand-Up & Walk & Sit-down & Stand-Up → Subject might be anxious or thinking
- Walk & Sit-down & Lay & Stand-up → Subject fall asleep and woke up, or took a nap, or felt dizzy and relaxed, based on Lay activity duration
- (Walk & Grasp & Cary) * 2 → Doing housework

10.7 Conclusions

We have presented a novel approach of human activity recognition in the continuously-evolving fields of computer vision and AI. Our system includes various techniques for recognizing an activity, segmenting a sequence of activities into action primitives, and, finally, classifying activities as normal or abnormal.

The system-methodology proposed here can contribute to the implementation of real-life complex activity recognition systems to effectively identify simple activities, segment complex activities, and, with the use of an experts' knowledge, recognize and appropriately define the meaning of activity sequences depending on the area it is applied. Overall, the tool does not only recognize activity, but it also offers the element of automatically interpreting sequences of activities.

References

1. L. Xia, C.-C. Chen, J.K. Aggarwal, *View Invariant Human Action Recognition Using Histograms of 3D Joints* (2012), pp. 20–27
2. S. Foix, G. Alenya, C. Torras, Lock-in Time-of-Flight (ToF) cameras: a survey. IEEE Sens. J. **11**(9), 1917–1926 (2011). https://doi.org/10.1109/JSEN.2010.2101060
3. W. Su, H. Eichi, W. Zeng, M. Chow, A survey on the electrification of transportation in a smart grid environment. IEEE Trans. Indus. Inf. **8**(1), 1–10 (2012). https://doi.org/10.1109/TII.2011.2172454
4. P. Rashidi, A. Mihailidis, A survey on ambient-assisted living tools for older adults. IEEE J. Biomed. Health Inform. **17**(3), 579–590 (2013). https://doi.org/10.1109/JBHI.2012.2234129
5. C. Urmson, W. Whittaker, Self-driving cars and the urban challenge. IEEE Intell. Syst. **23**(2), 66–68 (2008). https://doi.org/10.1109/mis.2008.34
6. A. Angeleas, N. Bourbakis, G. Tsihrintzis, Categorization of research surveys and reviews on human activities, in *2016 7th International Conference on Information, Intelligence, Systems & Applications (IISA)*, (Chalkidiki, 2016), pp. 1–6. https://doi.org/10.1109/iisa.2016.7785369
7. David B. Kaber, Mica R. Endsley, The effects of level of automation and adaptive automation on human performance, situation awareness and workload in a dynamic control task. Theor. Issues Ergon. Sci. **5**(2), 113–153 (2004)
8. A. Jaimes, N. Sebe, Multimodal human–computer interaction: a survey. Comput. Vis. Image Under. **108**(1–2), 116–134 (2007)
9. C. Hawes, C.D. Phillips, M. Rose, S. Holan, M. Sherman, A national survey of assisted living facilities. The Gerontologist **43**(6), 875–882 (2003)
10. S.S. Rautaray, A. Agrawal, Vision based hand gesture recognition for human computer interaction: a survey. Artif. Intell. Rev. **43**(1), 1–54 (2015)
11. S. Thompson, C. Bracken-Roche, Understanding public opinion of UAVs in Canada: a 2014 analysis of survey data and its policy implications. J. Unmanned Vehicle Syst. **3**(4), 156–175 (2015)
12. Á. Rocha and Asociación Ibérica de Sistemas y Tecnologías de Información, *2010 5th Iberian Conference on Information Systems and Technologies (CISTI 2010)* Santiago de Compostela, Spain, 16–19 June 2010 ; [including workshop papers] (Piscataway, NJ: IEEE, 2010), p. 226
13. L. Atzori, A. Iera, G. Morabito, The internet of things: a survey. Comput. Netw. **54**(15), 2787–2805 (2010)
14. ARCS, M. Beigl, F.J. Cazorla-Almeida, (Eds.), *Workshop proceedings/23th International Conference on Architecture of Computing Systems, ARCS 2010: February 22–23, 2010, Hannover, Germany* (VDE-Verl, Berlin, 2010)
15. M. Valera, S.A. Velastin, Intelligent distributed surveillance systems: a review. IEE Proc. Vision, Image, Signal Process. **152**(2), 192 (2005)
16. J.D. Culler, K. Lamb (eds.), *Just Being Difficult? Academic Writing in the Public Arena* (Stanford Univ. Press, Stanford, Calif, 2003)
17. R. Furman, J.T. Kinn, *Practical Tips for Publishing Scholarly Articles: Writing and Publishing in the Helping Professions*, 2nd edn. (Lyceum Books, Chicago, Ill, 2012)
18. C.H. Lim, E. Vats, C.S. Chan, Fuzzy human motion analysis: a review. Pattern Recogn. **48**(5), 1773–1796 (2015)
19. M.B. Holte, C. Tran, M.M. Trivedi, T.B. Moeslund, Human pose estimation and activity recognition from multi-view videos: comparative explorations of recent developments. IEEE J. Select. Top. Signal Process. **6**(5), 538–552 (2012)
20. J.K. Aggarwal, M.S. Ryoo, Human activity analysis: a review. ACM Comput. Surv. **43**(3), 1–43 (2011)
21. D. Das Dawn, S.H. Shaikh, A comprehensive survey of human action recognition with spatio-temporal interest point (STIP) detector. Vis. Comput. **32**(3), 289–306 (2016)
22. D.T. Nguyen, W. Li, P.O. Ogunbona, Human detection from images and videos: a survey. Pattern Recogn. **51**, 148–175 (2016)

23. S. Azary, A. Savakis, 3D action classification using sparse spatio-temporal feature representations, in *Advances in Visual Computing*, vol. 7432, ed. by G. Bebis, R. Boyle, B. Parvin, D. Koracin, C. Fowlkes, S. Wang, M.-H. Choi, S. Mantler, J. Schulze, D. Acevedo, K. Mueller, M. Papka (Springer Berlin Heidelberg, Berlin, Heidelberg, 2012), pp. 166–175
24. D. Weinland, R. Ronfard, E. Boyer, A survey of vision-based methods for action representation, segmentation and recognition. Comput. Vis. Image Underst. **115**(2), 224–241 (2011)
25. T.B. Moeslund, E. Granum, A survey of computer vision-based human motion capture. Comput. Vis. Image Underst. **81**(3), 231–268 (2001)
26. R. Poppe, A survey on vision-based human action recognition. Image Vis. Comput. **28**(6), 976–990 (2010)
27. Š. Obdržálek et al., Accuracy and robustness of Kinect pose estimation in the context of coaching of elderly population, in *2012 Annual International Conference of the IEEE Engineering in Medicine and Biology Society* (San Diego, CA, 2012), pp. 1188–1193. https://doi.org/10.1109/embc.2012.6346149
28. A. Angeleas, N. Bourbakis, Neural networks as classification mechanisms of complex human activities. Int. J. Artif. Intell. Tools **29**(05), 2050011 (2020)
29. A. Angeleas, N. Bourbakis, A two formal languages based model for representing human activities, in *2016 IEEE 28th International Conference on Tools with Artificial Intelligence (ICTAI)* (IEEE, 2016)
30. A. Angeleas, *A Multi-Formal Languages Collaborative Scheme for Complex Human Activity Recognition and Behavioral Patterns Extraction* (Wright State University, Diss, 2018)
31. L. Xia, C. Chen, J.K. Aggarwal, Human detection using depth information by Kinect, in *CVPR 2011 WORKSHOPS* (Colorado Springs, CO, 2011), pp. 15–22. https://doi.org/10.1109/cvprw.2011.5981811
32. M. Rahman, *Beginning Microsoft Kinect for Windows SDK 2.0: Motion and Depth Sensing for Natural User Interfaces* (Apress, 2017)
33. O. Masoud, N. Papanikolopoulos, Recognizing human activities, in *Proceedings of the IEEE Conference on Advanced Video and Signal Based Surveillance, 2003* (Miami, FL, USA, 2003), pp. 157–162. https://doi.org/10.1109/avss.2003.1217916
34. J. Ben-Arie, Z. Wang, P. Pandit, S. Rajaram, Human activity recognition using multidimensional indexing. IEEE Trans. Pattern Anal. Mach. Intell. **24**(8), 1091–1104 (2002). https://doi.org/10.1109/tpami.2002.1023805
35. F. Niu, M. Abdel-Mottaleb, View-invariant human activity recognition based on shape and motion features, in *IEEE Sixth International Symposium on Multimedia Software Engineering, Miami, FL, USA* (2004), pp. 546–556. https://doi.org/10.1109/mmse.2004.88
36. R. Hamid, Y. Huang, I. Essa, ARGMode—Activity Recognition using Graphical Models, in *2003 Conference on Computer Vision and Pattern Recognition Workshop, Madison, Wisconsin, USA* (2003), pp. 38–38. https://doi.org/10.1109/cvprw.2003.10039
37. A. Bhattacharjee et al., Stratum: a serverless framework for the lifecycle management of machine learning-based data analytics tasks. in *2019 {USENIX} Conference on Operational Machine Learning (OpML 19)* (2019)
38. Z. Li, B. Peng, Q. Huang, C. Weng, Karst: Transactional data ingestion without blocking on a scalable architecture. IEEE Trans. Knowl. Data Eng. https://doi.org/10.1109/tkde.2020.3011510
39. D.-S. Pham et al., Anomaly detection in large-scale data stream networks. Data Mining Knowl. Discov. **28**(1), 145–189 (2014)
40. S.C. Tan, K.M. Ting, T.F. Liu, Fast anomaly detection for streaming data, in *Twenty-Second International Joint Conference on Artificial Intelligence* (2011)
41. S.B. Gokturk, H. Yalcin, C. Bamji, A time-of-flight depth sensor—system description, issues and solutions (2004), pp. 35–35
42. H.H. Ali, H.M. Moftah, A.A.A. Youssif, Depth-based human activity recognition: a comparative perspective study on feature extraction. Future Comput. Inform. J. **3**(1), 51–67 (2018)
43. M. Panzner, P. Cimiano, Comparing hidden markov models and long short term memory neural networks for learning action representations, in *Machine 239 Learning, Optimization,*

and Big Data, vol. 10122, ed. by P.M. Pardalos, P. Conca, G. Giuffrida, G. Nicosia (Springer International Publishing, Cham, 2016), pp. 94–105

44. F. Karim, S. Majumdar, H. Darabi, S. Harford, Multivariate LSTM-FCNs for time series classification
45. J. Bergstra, Y. Bengio, Random search for hyper-parameter optimization. J. Mach. Learn. Res. **13**(2) (2012)
46. D.M. Hawkins, The problem of overfitting. J. Chem. Inf. Comput. Sci. **44**(1), 1–12 (2004)
47. R.C. Staudemeyer, E. Rothstein Morris, Understanding LSTM–a tutorial into long short-term memory recurrent neural networks. arXiv preprint arXiv:1909.09586 (2019)
48. J. Wu, *Introduction to Convolutional Neural Networks. National Key Lab for Novel Software Technology* (Nanjing University, China, 2017), p. 23
49. S. Yu, J. Ma, W. Wang, Deep learning tutorial for denoising. Electr. Eng. Syst. Sci. (2018)
50. S. McCann, J. Reesman, Object detection using convolutional neural networks (2015)
51. I. Goodfellow, *Deep Learning (Adaptive Computation and Machine Learning Series)* (Almohreraladbi, 2016)
52. X. Liu et al., Multi-task deep neural networks for natural language understanding. arXiv preprint arXiv:1901.11504 (2019)
53. D.W. Otter, J.R. Medina, J.K. Kalita, A survey of the usages of deep learning for natural language processing. IEEE Trans. Neural Netw. Learn. Syst. (2020)
54. D. Rao, B. McMahan, Natural Language Processing with PyTorch: Build Intelligent Language Applications Using Deep Learning (O'Reilly Media, Inc., 2019)
55. M. Oquab, L. Bottou, I. Laptev, J. Sivic, *Learning and Transferring Mid-level Image Representations Using Convolutional Neural Networks* (2014), pp. 1717–1724
56. B. Mahasseni, S. Todorovic, *Regularizing Long Short Term Memory with 3D Human-Skeleton Sequences for Action Recognition* (2016), pp. 3054–3062
57. B. Zhang, L. Wang, Z. Wang, Y. Qiao, H. Wang, *Real-Time Action Recognition with Enhanced Motion Vector CNNs*, (2016), pp. 2718–2726
58. L. Wang et al., Temporal segment networks: towards good practices for deep action recognition, in *Computer vision—ECCV 2016*, vol. 9912, ed. by B. Leibe, j Matas, N. Sebe, M. Welling (Springer International Publishing, Cham, 2016), pp. 20–36
59. C. Zhu, W. Sheng, Realtime recognition of complex human daily activities using human motion and location data. IEEE Trans. Biomed. Eng. **59**(9), 2422–2430 (2012). https://doi.org/10.1109/TBME.2012.2190602
60. X. Sun, C.-W. Chen, B.S. Manjunath, Probabilistic motion parameter models for human activity recognition, in *Object Recognition Supported by user interaction for service robots, Quebec City, Quebec, Canada*, vol. 1 (2002), pp. 443–446. https://doi.org/10.1109/icpr.2002.1044751
61. M.S. Ryoo, J.K. Aggarwal, Spatio-temporal relationship match: Video structure comparison for recognition of complex human activities, in *2009 IEEE 12th International Conference on Computer Vision, Kyoto* (2009), pp. 1593–1600. https://doi.org/10.1109/iccv.2009.5459361
62. M. Leo, T. D'Orazio, I. Gnoni, P. Spagnolo, A. Distante, Complex human activity recognition for monitoring wide outdoor environments, in *Proceedings of the 17th International Conference on Pattern Recognition*, vol. 4 (ICPR 2004, Cambridge, 2004), pp. 913–916. https://doi.org/10.1109/icpr.2004.1333921
63. L. Xia, C. Chen, J.K. Aggarwal, View invariant human action recognition using histograms of 3D joints, in *2012 IEEE Computer Society Conference on Computer Vision and Pattern Recognition Workshops, Providence, RI* (2012), pp. 20–27. https://doi.org/10.1109/cvprw.2012.6239233
64. M. Glykas, ed., *Fuzzy Cognitive Maps: Advances in Theory, Methodologies, Tools and Applications*, vol. 247 (Springer Science & Business Media, 2010)
65. A.J. Jetter, K. Kok, Fuzzy cognitive maps for futures studies—a methodological assessment of concepts and methods. Futures **61**, 45–57 (2014)

Chapter 11
A CU Depth Prediction Model Based on Pre-trained Convolutional Neural Network for HEVC Intra Encoding Complexity Reduction

Jiaming Li, Ming Yang, Ying Xie, and Zhigang Li

Abstract High efficiency Video Coding (HEVC) is the successor video compression standard of Advanced Video Coding (AVC), which achieves up to 50% higher compression ratio while maintaining the same video quality. One of the advanced techniques that HEVC adopts is dividing the input signal into Coding Units (CU) with various sizes, which ensures bitrate reduction while preserving visual details. The process that determines the size of CU is known as the Coding Tree Unit (CTU) partition. In order to determine the optimal CTU split pattern, the traditional approach uses exhaustive search to calculate the rate-distortion (RD) cost for every possible pattern, which is very computationally expensive and makes it impossible for real-time processing. In this paper, we present a hierarchical CU depth prediction model based on a pre-trained convolutional neural network to predict the CTU split pattern based on the image block. The model is built on a three-level classifier that can determine the depth of CU on each level and finally generate the overall partition of CTU. The experiments show that the model has the ability to reduce the intra-model encoding time by up to 73.80% with an average Bjøntegaard delta bitrate increase of 2.645%, which is a significant improvement over the encoding efficiency of HEVC.

Keywords High efficiency video coding · Coding tree unit partition · Convolutional neural network

J. Li · M. Yang (✉) · Y. Xie · Z. Li
College of Computing and Software Engineering, Kennesaw State University, Marietta, GA 30060, USA
e-mail: myang3@kennesaw.edu

J. Li
e-mail: jli36@students.kennesaw.edu

Y. Xie
e-mail: yxie2@kennesaw.edu

Z. Li
e-mail: zli8@kennesaw.edu

G. A. Tsihrintzis et al. (eds.), *Advances in Machine Learning/Deep Learning-based Technologies*, Learning and Analytics in Intelligent Systems 23,
https://doi.org/10.1007/978-3-030-76794-5_11

11.1 Introduction

With the development of communication technology, the data transmission bandwidth is constantly improving. However, it is still difficult to meet people's demand for real-time high-resolution video transmission. Video compression techniques can solve this problem by representing video data in a more compact and robust way to reduce the cost of storage and transmission. High Efficiency Video Coding (HEVC) is one of the widely used video compression standards that was accepted as the new video coding standard of the ITU-T Video Coding Experts Group and the ISO/IEC Moving Picture Experts Group in 2013. Compared to its predecessor H.264/AVC, it has the ability to reduce half of the bit rate without sacrificing the video quality and it supports high-definition video with resolution up to 8192 × 4320 (8 K). Even though HEVC has made a great achievement on bit-rate reduction, the encoding complexity is also dramatically increased. According to the study of Corrêa et al. [1], the average computational complexities for HEVC increased by 502.2% over H.264 for processing a video of 2560 × 1600 (2 K) resolution.

Coding Tree Unit (CTU) Partitioning is one of the techniques adopted by H.265/HEVC standard. CTU is the basic processing unit in H.265/HEVC, corresponding to the Macro Block (MB) in H.264/AVC. An MB usually has a fixed size of 16 × 16, while the size of CTU might be 16 × 16, 32 × 32, or 64 × 64. In addition, a CTU can be further partitioned recursively into multiple coding units (CU) following a quad-tree structure with depth from 0 to 3. Therefore, when the size of CTU is 64, the partitioning of CTU can be very complex. To find the optimal split pattern, the HEVC Test Model (HM, the reference software of HEVC) uses an exhaustive search to achieve rate-distortion optimization (RDO). It recursively computes the rate-distortion cost for each possible pattern to find the optimal partition. For a 4 K UHD video input with a resolution of 3840 × 2160, one single frame can be divided into at least 2,040 CTUs with the size of 64 × 64. Each CTU requires $1 + 4 + 16 + 64 = 85$ times of RDO cost calculation, and the total computational cost for this process makes it infeasible for real-time applications. The study of [2] reveals that the CTU partitioning procedure accounts for up to 90% of total computation in intra coding. Thus, CTU partition complexity reduction is a critical part to improve the efficiency of HEVC intra encoding.

Researchers have proposed a lot of solutions on CTU partition complexity reduction, and the methodologies can be classified into two categories: statistic-based methods and deep-learning-based methods. Generally, the statistic-based methods focus on the spatial distribution of CTUs to compress search space. For instance, Shang et al. [3] indicate that neighboring CUs usually contain similar textures and the optimal depth level of a CU has a strong correlation with its nearby CUs. Therefore, mode prediction in large CUs can be skipped early if the current CU contains complex texture information. Their algorithm achieves good compression quality, but the input frame must be preprocessed using a canny edge detector to determine the texture distribution, which is time-consuming. Also, the inherent parallelism of

CTU partitioning is degraded in this algorithm, since the depth search range of the current CU relies on its adjacent CUs.

Recently, as deep learning techniques reveal superiority in feature extraction and classification, researchers started to investigate CTU complexity reduction with the help of deep learning frameworks. In [2], Feng et al. proposed a CTU depth decision algorithm based on texture features and convolutional neural network (CNN) classification technique. They utilized CNN to reduce the search depth range of rate-distortion cost computation and were able to reduce the encoding time by an average of 27.54%. No auxiliary information is needed for this method, but even after depth range restriction, the algorithm still has overhead calculation complexity and it is still not efficient enough. Instead of focusing on search range reduction, Chen et al. [4] proposed two algorithms that use convolution neural networks for fast CU partition prediction and PU mode prediction respectively. Their proposed method was able to achieve 23.46% of time-saving for CU partition decision, with Bjontegaard delta bit rate (BDBR) increase of 0.08% and BDPSNR decrease of 0.06 dB. For PU mode prediction, their proposed method was able to save 23.61% of the encoding time.

Furthermore, another research group proposed a fast coding unit depth decision algorithm for intra coding of HEVC using an artificial neural network (ANN) and a support vector machine (SVM) [5]. In their research, machine learning is utilized to develop a fast algorithm for early CU splitting or termination to reduce intra coding computational complexity and time. Experimental results showed that the proposed fast algorithm saves at most 48.5% and on average 33% encoding time with a 1.55% BDBR loss compared to HM 15.0.

Song et al. reported a CNN-based arithmetic coding method for intra prediction modes in HEVC [6]. In their research, a customized CNN is used to predict the probability distribution of the intra prediction modes, and a multi-level arithmetic codec is adapted to compress the intra prediction modes with the predicted probability. Their results showed up to 9.9% bits saving compared with context-adaptive binary arithmetic coding (CABAC). In addition, some researchers attempted to use a convolutional neural network to predict CTU partition instead of exhaustive rate-distortion cost calculation. Kim et al. [7] designed a top-down neural-network architecture and trained it to predict whether or not to split CUs at each depth. For a given CU, the CNN model determines whether the CTU partition is supposed to stop at the current depth (NON-SPLIT) or to proceed to the lower depth (SPLIT). In the experiments, their algorithm reduced the computational overhead by 61.77% on average, and by a maximum of 73.45% with 3.91% Bjøntegaard-Difference bitrate (BD rate) degradation. Although there is a trade-off between encoding complexity and compression bitrate, their algorithm greatly improves the encoding efficiency. Li et al. [8] argued in their research that the CNN architecture of the previous study is relatively shallow, over-fitting is hard to avoid when the training data is not large enough. The model can hardly learn enough features. Thus, they explored a deep convolutional neural CNN model with various sizes of convolutional kernels and extensive trainable parameters. Their approach reduces intramode encoding time by 62.25 and 69.06%, with negligible BDBR of 2.12 and 1.38%. Their research demonstrated that a deeper network model performs better on CTU partitioning.

Inspired by pioneering researchers, we propose our CTU partition complexity reduction algorithm for HEVC intra encoding. The proposed method uses a hierarchical convolution neural network as a three-level CU depth classifier that makes predictions on the depth of CU on each level. Then the overall CTU partition pattern can be deduced by the depth of sub-CUs so that the overhead RD cost calculation can be substituted. First, training data is generated by HEVC reference software HM-16. The dataset contains YUV pixels value of CUs of size 64 × 64, 32 × 32, and 16 × 16 paired with their labels (depth). In consideration of the effect of Quantity Parameter (QP) on HEVC intra encoding, one CU is corresponding with 4 different labels under 4 QPs. Second, instead of designing new neural network architecture and training it from scratch, we adopted the methodology of transfer learning, which is using a pre-trained model for image classification as the convolutional base and fine-turning it to solve the task of CTU partition decision. Finally, we integrated our algorithm with HM-16. HM will skip RD cost calculation and exclusive quadtree search but assign the depth to CU directly according to the prediction of the proposed model.

11.2 H.265 High Efficiency Video Coding

Before stepping into the proposed methodology, it is necessary to reinforce the understanding of some critical terminologies of the H.265/HEVC standard.

11.2.1 Coding Tree Unit Partition

According to the HEVC standard, the input video signal (also called slice or frame) is firstly divided into equal size squares called CTUs. For a CTU with size $L \times L$, its length L can be chosen as $L = 16, 32$, or 64 based on the setting. HEVC then supports the partitioning of CTU into smaller CUs using a tree structure and quadtree-like signaling. This kind of flexible and recursive representation provides several benefits. The first one is the support of CU sizes greater than the conventional 16 × 16 size of H.264/AVC. As can be seen from Fig. 11.1, when the region is homogeneous, a large

Fig. 11.1 Processing units distribution of H.264/AVC (left) and H.265/HEVC (right)

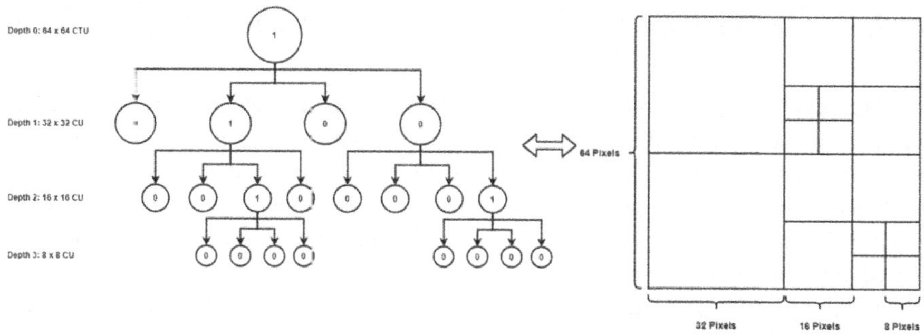

Fig. 11.2 QuadTree representation of CTU splitting pattern

CU can represent the region with a smaller number of symbols, so that encoding efficiency is improved compared to H.264/AVC [9].

A CU can be regarded as a $L \times L$ image area that can be further divided into 4 equal-size $\frac{L}{2} \times \frac{L}{2}$ smaller CUs. The structure of CUs can be represented by a Quadtree structure as illustrated in Fig. 11.2. A $\frac{L}{2} \times \frac{L}{2}$ CU can be further split as long as its length is greater or equal to 8. In other words, if we treat a 64×64 CTU as the root of the quadtree, the depth of the quadtree can range from 0 to 3. As indicated by Fig. 11.2, the quadtree representation of the CTU partition is denoted on the left side. The number "0" indicates current CU cannot be split (leaf), whereas "1" demonstrates that this CU can be expanded. On the right side of Fig. 11.2, the largest square represents a 64×64 CTU, or Largest Coding Unit (LCU). It was recursively divided into multiple CU with the smallest size 8×8.

11.2.2 Rate Distortion Optimization

The final CUs distribution is determined by Rate Distortion Optimization (RDO) using an exhaustive search on a full quadtree structure. Begin with the LCU and from top to the bottom, the algorithm adopted by HM will first calculate the RD cost of the whole LCU, then calculate the sum of RD cost of its sub-CUs. For each sub-CUs, this process will be done recursively until the length of sub-CU reaches 8. After that, the CTU partition with minimum total RD cost will be selected. Rate Distortion Optimization indeed brings optimization partition with minimum RD cost, but this method is not suitable for a time-sensitive application. A more efficient strategy is deemed necessary to reduce the CTU partition complexity with negligible rate-distortion loss.

11.2.3 CU Partition and Image Texture Features

Researchers have determined that the depth of CU is correlated with its surrounding texture features. Zhang et al.'s research revealed that the texture region of a frame tends to be coded with fine CU partition, while the homogeneous region is likely to be coded with coarse CU partition [10]. Therefore, they developed a gradient-based algorithm to accelerate prediction unit size and model. Similarly, the texture features can also be utilized to predict the depth of CU. For example, Fig. 11.3 shows the CTU partition of a portrait. The homogeneous region (e.g., background, forehead, and face) tends to have smaller depth; while the rich-texture areas (e.g. head, eye, and lip) tend to be divided into smaller sub-CUs.

Under the promise of the relationship between CU partition and texture features, if we train the deep learning model to learn the rule of CTU partition, the conventional brute-force search can be avoided so that HEVC intra encoding can be sped up.

11.3 Proposed Methodology

As we discussed earlier, predicting the depth range of CTU is not efficient enough because RD cost check is still unavoidable [3]. So, our proposed methodology is aimed to predict the CTU overall partition pattern directly.

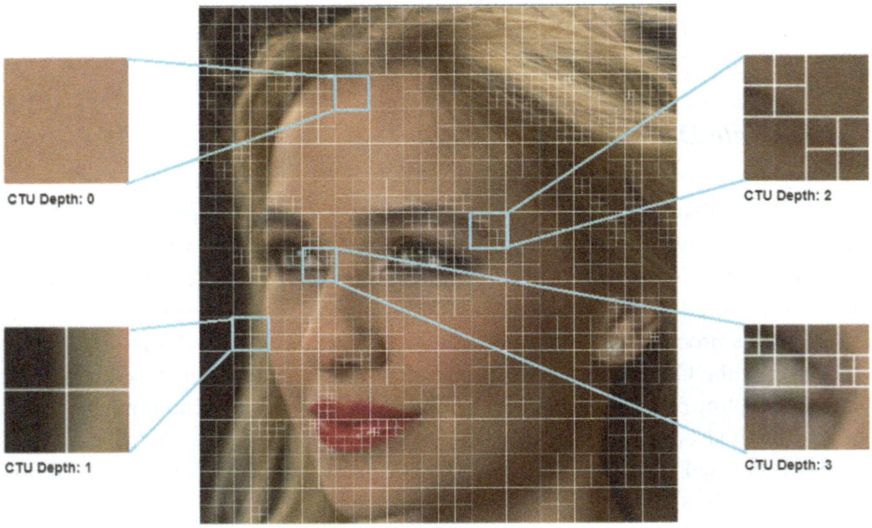

Fig. 11.3 CTUs with various maximum depth

11.3.1 The Hierarchical Classifier

For a 64×64 CTU, there can be as many as 83,522 different partition patterns [8]. Since there are too many possible patterns, it is difficult to train the model to give an immediate prediction. Thus, a combination of hierarchical models is preferred for this task.

Zhang et al. [5] presented a CU depth-decision algorithm using a three-level classifier. The procedure of the CTU partition prediction algorithm is similar to Depth-First Search (DFS). The top-level classifier is the CNN model that determines if the given LCU should be split on depth 0. If the decision is NON-SPLIT, then no further operation is performed at a lower depth. The algorithm should start to process the next CTU. But if the decision is SPLIT, then the LCU will be divided into four $\frac{L}{2} \times \frac{L}{2}$ smaller sub-CUs, and the algorithm goes to a lower depth. For each sub-CU, there is another CNN model that was trained to learn if a $\frac{L}{2} \times \frac{L}{2}$ CU should be split on depth 1. Similar to the procedure of depth 0, if the prediction is NON-SPLIT, the algorithm should stop here and start to process the next CU. Otherwise, it divides current CU into smaller CUs with length $\frac{L}{4} \times \frac{L}{4}$ until there is no more division necessary for the current CU.

While this design shows excellent performance on CTU pattern prediction, the ability of CNN to extract deep and abstract features has not been fully utilized. In their work, each depth is a binary classification problem. But in reality, CNN could handle more complex classification tasks. Therefore, we modified their algorithm and transformed the two-level model into a ternary classifier. Unlike their model that predicts SPLIT and NON-SPLIT, our model will make a prediction on the depth of CU from "0" to "2". If the prediction shows the depth of CU is 0, then it is equal to NON-SPLIT. But if the depth is "1", then the current CU will be split into four sub-CUs. There is no need to check four sub-CUs because we know the depth of CU is "1", its sub-CUs cannot be further split. We only need to check every sub-CU when the prediction is "2" ("2" means current CU's depth is greater or equal to 2. In this case, we know that at least one sub-CU needs to be split, but we do not know which one should be split, so the checking is necessary. As a result, our model further improves the encoding efficiency by avoiding needless checking. Figure 11.4 shows the structure of our improved classifier.

11.3.2 The Methodology of Transfer Learning

The author of [5] chose to design and train a convolutional neural network from scratch. However, there might be a more efficient way to build the model by adopting the methodology of transfer learning. Transfer learning is the method that draws knowledge from other related tasks to assist with current task learning. In this scenario, we can use the pre-trained model as a feature extractor to extract the texture features of CUs and add our own dense layer as a classifier to decide the depth of CUs.

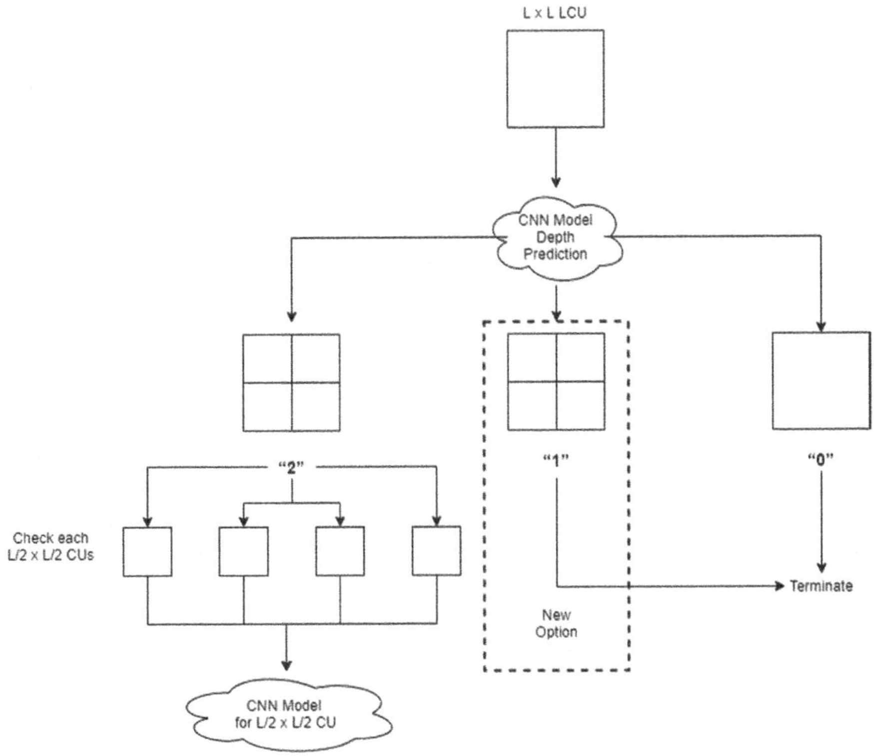

Fig. 11.4 Diagram of improved classifier

A pre-trained model is a model that has already been fully trained in a huge dataset and can yield good classification results. There are plenty of ready-to-use pre-trained models that we can utilize, for example, VGG [11], Densnet [12], and MobileNet [13]. Those models are provided along with their weights trained on the ImageNet dataset, which is a large image dataset organized according to the WordNet hierarchy with more than 14 million images included. Most pre-trained models have more than 70% Top-1 accuracy and 90% Top-5 accuracy for 1000 classes which means those models can extract deeper and more abstract features of image data. Similarly, CTU splitting is based on the block texture, so we can use pre-trained models to assist with our task.

There are mainly two advantages to using a pre-trained model. First, it eliminates the need for a huge dataset to train the model. Since the pre-trained model is already well trained and can yield good results on image classification, we only need to train it using a relatively small amount of data to discriminate three different kinds of CUs. Second, the training time will be dramatically reduced if we train from a pre-trained model. Most pre-trained models have very deep structures, but the fine-tuning only requires us to train part of its network. For most of the layers, their weights will stay unchanged during the training. Just a few layers will need to be tuned when we train the model.

11.3.3 Structure of Convolutional Neural Network

The proposed convolutional neural network model serves as a single CU depth predictor. By combining three models together and collecting the depth of each individual CUs, the overall partition pattern can be predicted directly to reduce HEVC encoding complexity. The proposed model can be divided into three parts: input layers, convolutional base, and fully connected layers.

The input layer is the image block of CU. Since we designed a three-level hierarchical classifier, there are three models for one QP, we refer to them as Model-64, Model-32, Model-16 respectively according to the shape of their input layer.

The second part of the model is the convolutional layers of a pre-trained model and the weights, we refer to it as the convolutional base. These convolutional layers were trained to extract deep and abstract features and achieve good accuracy. Normally the pre-trained models are very deep, but as we discussed earlier, only a few convolutional layers need to be tuned during the training, and other layers will stay unchanged. Therefore, it is not difficult to train the model despite the depth of the overall network. Model-16 is the special case here. Since the size of 16×16 CU is relatively small, there is no need to use deep convolutional layers, and the model is able to provide a good result. For the convolutional layers of Model-16, the size of kernels is set to 3×3, and the stride is set to 1 in both x and y directions.

Three fully connected (FC) layers are activated by the ReLU function. They serve as feature classifier. The feature maps generated by the convolutional layers will be flattened to a one-dimensional vector and then passed to the FC layers. Most pre-trained models focus on image classification, so we modified its original fully connected layers and trained them to predict CU depth based on the features extracted by preceding convolutional layers. Due to the limited size of training data, dropout layers are added between the FC layer to prevent over-fitting. The drop out ratio is set to 0.5, which means 50% units in FC layers will be randomly set to 0. Thus, the model will not overly rely on the local feature and its generalization ability will be improved. Figure 11.5 depicts the detailed structure of Model-64, Model-32, and Model-16.

Plenty of pre-trained models exist and most of them perform excellently on the task of image classification. However, some models might overlook the features we need for the task of CU depth prediction. Therefore, we evaluated several classic pre-trained models and it is found that VGG16 yielded the best test accuracy. Theoretically, a higher accuracy will get better BD-BR and BD-PSNR because the test accuracy indicates the similarity of the ground truth. Therefore, we chose VGG-16 as the convolutional base of our model to gain the best performance.

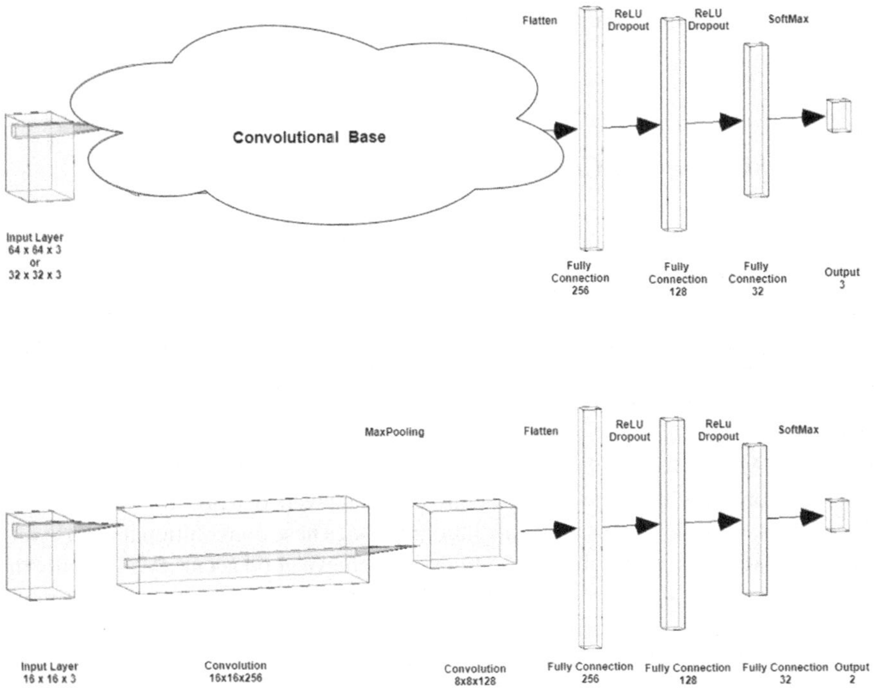

Fig. 11.5 Structure of proposed model (up: model64 and model32; down: model16)

11.3.4 Dataset Construction

In order to train the neural networks to learn the texture features of a CU and to predict the depth of it, a dataset should be created. The dataset should contain the image block of CUs paired with their depth as the label. We first downloaded several YUV raw videos from the UVG dataset [11] and Xiph.org, and then classified them into two groups according to their display resolution (some videos are resized). Table 11.1 displays the attributes of the videos we used to construct the dataset.

We used the reference software HM to encode the video sequences to obtain the ground truth depth information of CU. For a 64×64 CTU, the complex CTU partition pattern can be easily represented by a 4×4 matrix, where each entry of the matrix represents the depth of an 8×8 CU, and the location of CU can be determined by the entry indices (Fig. 11.6).

The CTU depth matrix serves as a label, while the corresponding data is the image block of CTU. To make our dataset lightweight, we use a tuple (F, I) to index the CTU, where F is the frame number and I is the CTU index. The data along with labels will be stored in a pickle file.

To load the CUs and labels depicts the depth, first the frames need to be extracted from the video sequence with YUV 4:2:0 pixel format. Because the adjacent frames

Table 11.1 List of video sequence

Class	Video name	Display resolution	Total frame/frame rate
A	ReadySteadyGo	1080P	600/120
	YachtRide	1080P	600/120
	HoneyBee	1080P	600/120
	Jockey	1080P	600/120
B	OldTown	720P	500/20
	Hields	720P	504/50
	FlowerPan	720P	600/50
	Ducks	720P	500/20
	FlowerKids	720P	600/50
	RaceNight	720P	600/50

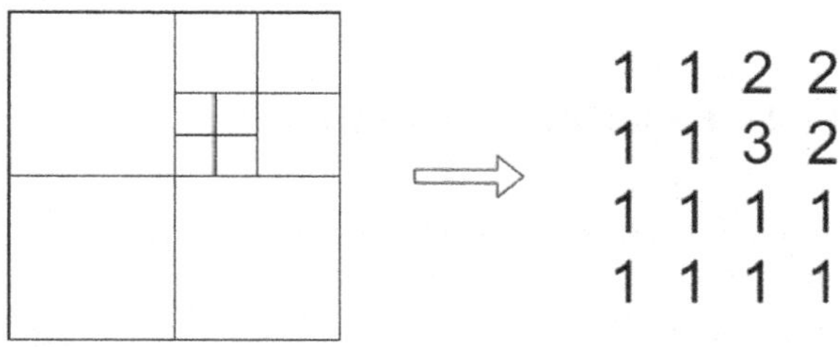

Fig. 11.6 Matrix representation of CTU partition pattern

have high continuity, not the CUs of every frame are added to the dataset to keep the diversity of the dataset and help the model to learn various features. For the selected frames, the frame number and CTU index helps us locate the image block of CTU. In addition, we can easily extract the single CU partition information by the depth matrix. For example, if the sum of the entries equals 0, then the 64×64 CU's label is "0" which means it cannot be split. If the sum is 16, then the label for this 64×64 CU is "1". Otherwise, we assign label "3" to this CU, and then divide this CU into four sub-CUs. Similarly, we divide the depth matrix into 4 sub-matrices. By doing the above procedure recursively, we can label every CU in a CTU.

Notice that with different Quantity Parameters (QP) settings, the CTU partition pattern tends to be different. In video encoding with higher QP, the depth of CTU becomes deeper because of the need to achieve higher compression efficiency [4]. If we do not consider the effect of QP, the neural network will yield wrong predictions and the RD-performance will be degraded. Therefore, for each video sequence, we created 4 sub-datasets that correspond to 4 different QPs, specifically, 22, 27, 32, and 37. Table 11.2 shows the detailed data and labels distribution of our dataset.

Table 11.2 Dataset distribution

QP	Size	Label: "0"	Label: "1"	Label: "2"
22	64 × 64	45,472	54,179	248,989
	32 × 32	271,799	98,763	625,394
	16 × 16	782,843	1,718,733	N/A
27	64 × 64	55,734	61,986	23,0920
	32 × 32	280,092	10,9015	53,4573
	16 × 16	793,366	1,344,926	N/A
32	64 × 64	69,005	60,385	219,250
	32 × 32	291,768	124,795	460,437
	16 × 16	773,504	7,068,244	N/A
37	64 × 64	83,795	64,028	200,817
	32 × 32	301,360	144,240	357,668
	16 × 16	681,426	749,246	N/A

11.4 Experiments and Results

In this section, we will first find the optimal pre-trained model for this task, then utilize it as the convolutional base of our model to evaluate the performance of the proposed methodology. Specifically, we will integrate the proposed model to HM, and the Bjøntegaard delta bitrate (BD-BR), Bjøntegaard delta PSNR (BD-PSNR) of modified HM will be compared with the original HM.

There is a total of 10 video sequences in two categories in our dataset as shown in Table 11.1. For each experiment, one video will be picked up as test data, other videos (dataset) in the same class will be used to train the model, which means the weight of the model will be updated through backpropagation except the layers that have been frozen. Test data is only used to evaluate the performance of the model by test accuracy and loss value. Test accuracy and loss will not affect the weights of the model, and the training data and test data are separate during our experiment. The reason we introduce test data in training is that we can avoid over-fitting by viewing the changes on test loss. Fine-tuning is necessary if the test loss keeps increasing.

For each QP in {22, 27, 32, 37}, there are 3 models that predict the depth of CU with dimensions 64 × 64, 32 × 32, and 16 × 16, respectively. Therefore, a total of 12 models are trained for each test data. The batch size is set to 256, 512, 1024 for Model-64, Model-32, and Model-16. The models are all compiled with Adma optimizer with an initial learning rate of 0.001. We also use a learning rate regulator to adjust the learning rate during the runtime. If the validation loss stops decreasing for 3 epochs, the learning rate will decrease by 20%. The learning rate regulator helps prevent overfitting. Since we train the model based on the pre-trained model, the model only has to be iterated for few epochs. In our experiment, Model64 and Model32 will be trained for 10 epochs, whereas Model16 needs 20 epochs to be fully trained. Our experiment indicated that, after the training, most models can predict the depth for one specific CU with over 80% accuracy. Figure 11.7 illustrates the training and test accuracy of each level's model with 4 different QPs.

The configuration file *encoder_intr_main.cfg* provided by HM is designed for HEVC intra prediction, so it will be used for parameter setting. Only the value of QP will be modified to collect different data. Other parameters in the configuration file will remain unchanged. Test videos are encoded by both original HM and our modified HM for four different QP in {22, 27, 32, 37}. After HM finishes encoding the specified frames, a summary report that contains bitrate information, PSNR on each channel, and total encoding time will be generated automatically by HM. Next, according to the bitrate and the YUV-PSNR for the same input video sequence of four different QPs, the RD performance can be evaluated by calculating BD-BR and BD-PSNR. Table 11.3 shows the BD-BR increment and BD-PSNR reduction of our method compared with the original HM. We can see from Table 11.3 that the BD-BR increment of our approach is 2.64%. In addition, our approach also results

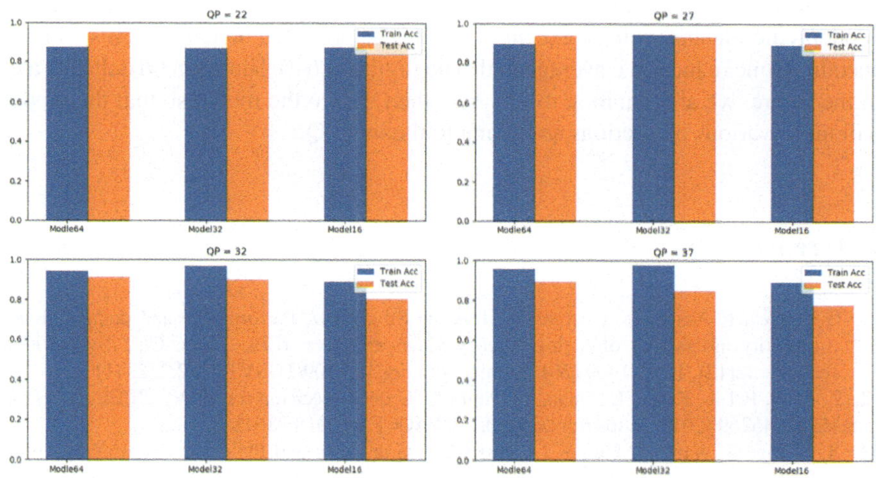

Fig. 11.7 Train and test accuracy for test video sequence "FlowerPan"

Table 11.3 BD-BR, BD-RSNR, and time reduction of proposed method

Video name	BD-BR (%)	BD-RSNR	Time reduction				
			QP = 22 (%)	QP = 27 (%)	QP = 32 (%)	QP = 37 (%)	Average (%)
Ducks	2.62	−0.13	73.80	70.05	67.65	64.52	69.01
FlowerKids	1.92	−0.10	50.00	45.16	41.38	37.03	43.39
FlowerPan	2.44	−0.14	51.16	46.15	38.23	32.25	41.95
OldTown	2.64	−0.11	63.88	52.94	43.33	37.03	49.30
RaceNight	3.23	−0.14	63.64	63.33	60.71	59.26	61.74
Hields	3.02	−0.15	40.47	35.13	21.21	13.33	27.53

in 0.13 dB PSNR degradation. In return, the proposed method reduces the encoding time by 48.82% on average. Our approach performs very well as can be seen from the experimental results. The pre-trained model plays an important role in our model. With its convolutional layers, the feature of CU is classified quickly, thus our model produced an excellent result on RD performance without training on a large-scale dataset for a long period of time.

11.5 Conclusion

As discussed earlier in this paper, the CTU split pattern prediction method in the HEVC reference software HM is not well-suited for real-time application scenarios. In this study, we take advantage of the convolutional neural network to improve the prediction speed of the CTU split pattern. Our results demonstrated that the pre-trained convolutional neural network can be used to handle the task of CU depth prediction and yield an excellent result. The proposed model achieves 48.82% encoding time reduction o average with a negligible 2.64% Bjøntegaard delta bitrate. In the future, we are planning to integrate the QP into the model so that the model will make various predictions according to the input QP.

References

1. G. Correa, P. Assuncao, L. Agostini, L.A. da Silva Cruz, Performance and computational complexity assessment of high-efficiency video encoders. IEEE Trans. Circ. Syst. Video Technol. **22**(12), 1899–1909 (2012). https://doi.org/10.1109/TCSVT.2012.2223411.
2. Z. Feng, P. Liu, K. Jia, K. Duan, Fast intra CTU depth decision for HEVC. IEEE Access **6**, 45262–45269 (2018). https://doi.org/10.1109/ACCESS.2018.2864881
3. X. Shang, G. Wang, T. Fan, Y. Li, Fast CU size decision and PU mode decision algorithm in HEVC intra coding, in *2015 IEEE International Conference on Image Processing (ICIP)* (2015), pp. 1593–1597. https://doi.org/10.1109/ICIP.2015.7351069.
4. K. Chen, X. Zeng, Y. Fan, CNN oriented fast CU partition decision and PU mode decision for HEVC intra encoding, in *2018 14th IEEE International Conference on Solid-State and Integrated Circuit Technology (ICSICT)* (2018), pp. 1–3. https://doi.org/10.1109/ICSICT.2018.8564981.
5. Y. Zhang, S. Kwong, X. Wang, H. Yuan, Z. Pan, L. Xu, Machine learning-based coding unit depth decisions for flexible complexity allocation in high efficiency video coding. IEEE Trans. Image Process. **24**(7), 2225–2238 (2015). https://doi.org/10.1109/TIP.2015.2417498
6. R. Song, D. Liu, H. Li, F. Wu, Neural network-based arithmetic coding of intra prediction modes in HEVC, in *2017 IEEE Vis. Commun. Image Process. VCIP* (2017), pp. 1–4. https://doi.org/10.1109/VCIP.2017.8305104.
7. K. Kim, W.W. Ro, Fast CU depth decision for HEVC using neural networks. IEEE Trans. Circ. Syst. Video Technol. **29**(5), 1462–1473 (2019). https://doi.org/10.1109/TCSVT.2018.2839113
8. T. Li, M. Xu, X. Deng, A deep convolutional neural network approach for complexity reduction on intra-mode HEVC, in *2017 IEEE International Conference on Multimedia and Expo (ICME)* (2017), pp. 1255–1260. https://doi.org/10.1109/ICME.2017.8019316.

9. I.-K. Kim, J. Min, T. Lee, W.-J. Han, J. Park, Block partitioning structure in the HEVC standard. IEEE Trans. Circ. Syst. Video Technol. **22**(12), 1697–1706 (2012). https://doi.org/10.1109/TCSVT.2012.2223011
10. Y. Zhang, Z. Li, B. Li, Gradient-based fast decision for intra prediction in HEVC, in *2012 Visual Communications and Image Processing* (2012), pp. 1–6. https://doi.org/10.1109/VCIP.2012.6410739.
11. K. Simonyan, A. Zisserman, Very deep convolutional networks for large-scale image recognition, *ArXiv14091556 Cs* (2015). Accessed: Aug. 26, 2020. [Online]. Available: http://arxiv.org/abs/1409.1556.
12. G. Huang, Z. Liu, L. van der Maaten, K.Q. Weinberger, Densely connected convolutional networks, *ArXiv160806993 Cs* (2018). Accessed: Aug. 26, 2020. [Online]. Available: http://arxiv.org/abs/1608.06993.
13. A.G. Howard et al., MobileNets: efficient convolutional neural networks for mobile vision applications, *ArXiv170404861 Cs* (2017). Accessed: Aug. 26, 2020. [Online]. Available: http://arxiv.org/abs/1704.04861.

The manufacturer's authorised representative in the EU is Springer Nature Customer Service Centre GmbH, Europaplatz 3, 69115 Heidelberg, Germany. If you have any concerns regarding our products, please contact ProductSafety@springernature.com

Printed and bound by CPI Group (UK) Ltd, Croydon, CR0 4YY
30/09/2026
02225287-0006